Elemental Narratives

AnthropoScene
THE SLSA BOOK SERIES

Lucinda Cole and Robert Markley, General Editors

Advisory Board:

Stacy Alaimo (University of Texas at Arlington)
Ron Broglio (Arizona State University)
Carol Colatrella (Georgia Institute of Technology)
Heidi Hutner (Stony Brook University)
Stephanie LeMenager (University of Oregon)
Christopher Morris (University of Texas at Arlington)
Laura Otis (Emory University)
Will Potter (Washington, D.C.)
Ronald Schleifer (University of Oklahoma)
Susan Squier (Pennsylvania State University)
Rajani Sudan (Southern Methodist University)
Kari Weil (Wesleyan University)

Published in collaboration with the Society for Literature, Science, and the Arts, AnthropoScene presents books that examine relationships and points of intersection among the natural, biological, and applied sciences and the literary, visual, and performing arts. Books in the series promote new kinds of cross-disciplinary thinking arising from the idea that humans are changing the planet and its environments in radical and irreversible ways.

Elemental Narratives

Reading Environmental
Entanglements in Modern Italy

Enrico Cesaretti

The Pennsylvania State
University Press
University Park,
Pennsylvania

Library of Congress Cataloging-in-Publication Data
Names: Cesaretti, Enrico, 1963– author.
Title: Elemental narratives : reading environmental entanglements in modern Italy / Enrico Cesaretti.
Other titles: AnthropoScene.
Description: University Park, Pennsylvania : The Pennsylvania State University Press, [2020] | Series: AnthropoScene: the SLSA book series | Includes bibliographical references and index. Summary: "Using an ecomaterialist conceptual framework, addresses interconnected stories from fiction, nonfiction, works of visual art, and physical sites in Italy and elsewhere"— Provided by publisher.
Identifiers: LCCN 2020035071 | ISBN 9780271087733 (hardback) | ISBN 9780271087740 (paper)
Subjects: LCSH: Ecology in literature. | Italian literature— 20th century— History and criticism. | Italian literature— 21st century— History and criticism. | Environmentalism in literature. | Pollution in literature.
Classification: LCC PQ4053.E26 C47 2020 | DDC 850.9/3553— dc23
LC record available at https://lccn.loc.gov/2020035071

Copyright © 2020 Enrico Cesaretti
All rights reserved
Printed in the United States of America
Published by The Pennsylvania State University Press,
University Park, PA 16802-1003

The Pennsylvania State University Press is a member of the Association of University Presses.

It is the policy of The Pennsylvania State University Press to use acid-free paper. Publications on uncoated stock satisfy the minimum requirements of American National Standard for Information Sciences— Permanence of Paper for Printed Library Material, ANSI Z39.48-1992.

Contents

Acknowledgments | vii

Introduction: Elemental Narratives | 1

1. Modernist Matters | 14

2. Slick Territories: Petroculture, Italian Style | 46

3. Apua Ma(t)ter: Narratives of Marble | 90

4. Steel and Asbestos: Stories of Toxic Lands and Bodies in Tuscany and Beyond | 124

5. Concrete and Asphalt: Geographies of Environmental Disruption in Modern Italy | 154

Epilogue | 201

Notes | 205

References | 225

Index | 245

Acknowledgments

Unsurprisingly, this book is the outcome of an entangled mix of my writing and the inspiration, conversations, exchanges of ideas, encouragement, support, and suggestions I received over recent years, in various guises, by a number of colleagues, students, and friends. With no hope of being exhaustive (and I apologize for any memory lapses), I would like to acknowledge a few people who stood out and helped me shape and complete this project.

First, I wish to express my gratitude to Serenella Iovino, whose own groundbreaking work (and in collaboration with Serpil Oppermann) opened up a whole new chapter in my research. Her work made my own book possible and continues to be an exemplary source of creative, cutting-edge scholarship. I am also indebted to Damiano Benvegnù, whose expert discussion of Primo Levi's *The Periodic Table* over coffee at the University of Virginia (UVA) "Corner" was among the initial sparks for this project, and to Elena Past, who patiently read a couple of my early chapters and provided insightful suggestions and feedback.

I am especially grateful to Monica Seger, who has always been an available, knowledgeable, and most generous interlocutor since the time we coorganized our first panel on Italian ecocriticism at an American Association for Italian Studies (AAIS) conference. Among the growing community of scholars working within the Italian-Mediterranean environmental humanities whom I have been constantly learning from and interacting with at past AAIS and Association for the Study of Literature and Environment (ASLE) conferences or in other venues, I am thankful to Pasquale Verdicchio, Massimo Lollini, Marco Armiero, Stefania Barca, Matteo Gilebbi, Danila Cannamela, Serena Ferrando, Paolo Chirumbolo, Patrick Barron, Miranda Macphail, Ilaria Tabusso Marcyan, Emiliano Guaraldo, and Luca Bugnone. I also wish to thank all the colleagues and (by now ex-) graduate students who kindly gave me the opportunity to present parts of this project in various academic settings: Silvia Ross and Mark Chu at the University of Cork, Luca Somigli at the University of Toronto,

Hiromi Kaneda at Rutgers University, Francesco Bratos at the University of North Carolina at Chapel Hill, and Laura Di Bianco at Johns Hopkins University.

Being part of the 2016–17 Mellon Humanities Fellows Program thanks to an award granted by the UVA's Institute of Humanities and Global Cultures allowed me to organize a symposium on the "Environmental Post-humanities in the Anthropocene." In addition to some of the colleagues I already mentioned, I learned and benefited from rewarding chats with Karen Pinkus, Heather Sullivan, Federico Luisetti, and Louise Westling. My intellectual horizons have been expanding also thanks to the work of all colleagues affiliated with the Environmental Humanities group at UVA.

Research support in the form of a UVA Faculty Stipend for Summer Research (2017) and a UVA Sesquicentennial Associateship (spring 2020) helped me finalize this project, and I am therefore thankful to the dean of the UVA College of Arts and Sciences and to the vice president for Research and Graduate Studies.

I am obliged to Joseph Luzzi, who kindly helped me draft a better pitch for my proposal, and to Liz Launer at Speak! Language Center for carefully reading the draft of my manuscript, correcting and improving my English writing, and letting me know whenever she got lost in some never-ending paragraphs in my Italianate prose.

I wish to express my deepest gratitude to Lucinda Cole and Robert Markley, editors of the SLSA AnthropoScene Series, and to Kendra Boileau, Alex Vose, Megan Grande, and the whole editorial team at Penn State University Press for believing in this book and assisting me during all the phases of this long process. This book has also been undoubtedly improved because of the pointed critical comments and suggestions of the two (no longer) anonymous readers, to whom I am immensely thankful.

Finally, thanks to Christina, Mila, and Delta for bearing with me . . . and reminding me of what really matters.

The section "'A Life of Metal': An Ecocritical Reading of Silvia Avallone's *Swimming to Elba*" in chapter 4 was first published as "'A Life of Metal': An Ecocritical Reading of Silvia Avallone's *Acciaio*," *Ecozon@* 5, no. 2 (2014): 107–22 (Creative Commons license).

A shorter version of the section "Ecofuturism? Nature, Matter, and Body in F. T. Marinetti" in chapter 1 was first published as "Eco-futurism? Some

Thoughts on Nature, Matter, and Body in F. T. Marinetti," in *Modernism and the Avant-Garde Body in Spain and Italy*, ed. Nicolás Fernández-Medina and Maria Truglio (New York: Routledge, 2016), 232–48 (copyright 2016; reproduced in reviewed and expanded form by permission of Taylor and Francis Group, LLC, a division of Informa PLC).

Introduction
Elemental Narratives

Break the pact, rewrite the agreement:
to begin with, dress yourself
in hard bark, resonant metal:
be vegetable and many animals.

Rompi il patto, riscrivi l'intesa:
per cominciare, vestiti
di dura corteccia, di sonoro metallo:
sii vegetale e molti animali.

—Ermanno Krumm, *Animali e uomini*

Understanding matter is necessary to comprehend the universe and ourselves.

Comprendere la materia è necessario per comprendere l'universo e noi stessi.

—Primo Levi, *The Periodic Table*

Taking cues from specific chemical elements in Mendeleev's periodic table—from "Argon" to "Carbon," passing through "Hydrogen," "Nickel," "Arsenic," and so on—the twenty-one chapters forming Primo Levi's *The Periodic Table* have been described as "encounters with matter, seen sometimes as mother and sometimes as enemy" (Levi 1984, back jacket; incontri con la materia, vista volta a volta come madre o come nemica).[1] Well-known outside of Italy especially for his novels about surviving the Holocaust, Levi was also a professional organic chemist who, in this later book, weaves together the story of his life, one shared by an entire generation that resisted Nazi fascism and coped with World War II, and the story of the many challenges and satisfactions he derived from chemistry and writing.

As he imaginatively implies in the last chapter in the collection, "Carbon," these two disciplines—writing and chemistry—are fundamentally connected at the biological and atomic levels.[2] They can offer a glimpse of the relationships between the universe and human beings and, by extension, of the interwoven physical, chemical, spiritual, and moral dimensions underpinning them. Most relevant to the objectives of my book, Levi muses on the real and metaphorical correlations between the elemental and the human sphere and on their more or less hidden analogies in terms of activities and creative potential. By doing so, he also implicitly suggests—well ahead of his time and recent theoretical elaborations—that forms of agency and expressivity are not prerogatives of human beings but are shared by the materials with which we interact and that, in turn, interact with us.

For instance, when we read in the chapter "Lead" that this "is a metal which feels fatigued, perhaps it's tired of changing and no longer wishes to transform" (Levi 1984, 87; un metallo che senti stanco, forse stanco di trasformarsi e che non si vuole trasformare più) or, in "Mercury," that a character (Hendrik) "seems to have turned into mercury, [the metal] running in his veins and leaking out his eyes" (104; sembrava diventato mercurio, che gli corresse per le vene e gli trapelasse dagli occhi), we are not merely confronted with colorful, metaphorical prose describing processes of anthropomorphism and objectification. What emerges from these lines is also Levi's reconsideration of dualistic dichotomies and long-established boundaries and his desire to use his storytelling skills to convey the ontological proximity of the human and the nonhuman. As Jane Bennett puts it in *Vibrant Matter: A Political Ecology of Things*—a work that provides essential support for my book's conceptual framework—"In revealing similarities across categorical divides and lighting up parallels between material forms in 'nature' and those in 'culture,' anthropomorphism can reveal isomorphism" (2010, 99). In this sense, to self-servingly adapt the verses quoted in Krumm's epigraph to these circumstances, Levi indeed contributes to imaginatively "break the pact, / rewrite the agreement" that has long privileged a traditional, dualistic (Cartesian), and anthropocentric worldview and rather inserts the human in a mutually constructive and revelatory dialogue with the nonhuman and the inhuman.[3]

Situated at the juncture of Italian studies and ecocriticism, *Elemental Narratives: Reading Environmental Entanglements in Modern Italy* takes *The*

Periodic Table as an initial source of structural, thematic, and philosophical inspiration. Namely, not only does each of its five chapters revolve around one (or two) specific substance(s), but it also shares Levi's original vision of a reality in which the encounter of matters, storytelling, and human beings has profound cognitive, ethical, and political implications.

I suggest that engaging with a selection of "narratives of entanglement" that speak of our simultaneous (co)existence in both imaginative and material universes does not just promise to generate new knowledge and more participative, affective responses to environmental issues. As this gesture aims at complementing quantitative, data-based information and fact-finding, it also hopes to induce the systemic changes needed to seriously address existing ecological crises. More concretely, along with the several scholars who codrafted the "Humanities for the Environment: A Manifesto for Research and Action," I believe that insights from fiction and nonfiction, the arts, and other humanistic disciplines may modify perceptions and attitudes, increase our awareness and understanding, and in turn, incite at least some of us toward a more responsible, active, and "efficacious engagement with global environmental challenges" (Holm et al. 2015, 978).

Heeding the recent "turn to the material" in the environmental humanities and Serenella Iovino and Serpil Oppermann's invitation to consider "matter not solely as it appears in texts, but as a text itself" (2014, 6), I too seek to literally engage with matter. Namely, I ask what stories it tells, and lets us tell, in the wake of increasingly troubling predictions about the health of global environments, of multidisciplinary challenges to both the limits and boundaries of humans and things, and of reassessments of the separate nature of mind, body, language, and place.[4] Fully endorsing the view that matter is not just a background setting but a protagonist and an active agent in our becoming that "produces ongoing configurations of signs and meanings that can be interpreted as stories" (Oppermann 2017, 293), *Elemental Narratives* individuates and examines some of these stories as they emerge in different media, bodies, and places. It reflects not only on how human beings deal, and have historically dealt, with some of the substances that contribute to shaping their world but on how such materials, by interacting with human imagination, express their own stories and construct meaning, effecting changes in human aesthetic practices, knowledges, perspectives, and ways of being and behaving.

More specifically, my book addresses a selection of texts in an Italian-Mediterranean landscape—considered simultaneously in its cultural and topological, semiotic, and geographical dimensions—in which the boundaries between what is human and nonhuman, organic and inorganic tend to become blurred and indistinct. As such, these texts ultimately delineate an ontological condition of interconnectedness, reciprocity, and relationality among materials, places, and living organisms. Echoing Iovino's remarks about the appropriateness of considering the study of the dynamics of specific places "as generative of categories" that can be applied to understanding many other places, here Italy functions at once as territory and map, as a specific site of analysis and a cognitive instrument (2016a, 2–3). It is a fact that environmental issues and, by extension, ecocritical discourses do not have strict borders and ultimately are "'travelling theories' . . . irreducible to one geographical, national or methodological origin" (DeLoughrey and Handley 2011, 16). In this light, I consider Italy as a bioregional natural-cultural microcosm that despite its specific singularities (historical, cultural, geographic, and so on), still enlightens and is enlightened by the situation in the macrocosm of other places and collectives. Of course, as Erin James observes, "People around the world imagine, inhabit, and experience their environments differently," and Italians, in certain aspects (given, for instance, their strong humanistic heritage and historical relationship with the notion of landscape), may do that even more "differently" than others.[5] However, she also argues that "narratives, with their power to immerse readers in environments and environmental experiences different from their own, can reveal perceptual points of difference, clarify the interests of those who imagine and inhabit an environment in a specific way . . . they can open channels of communication . . . and encourage an environmental awareness that may help to craft more equitable, just, and nonpartisan environmental policies" (2015, 208).

If James's remarks make perfect sense in the context of the internationally produced, geographically expansive, yet also ultimately anglophone postcolonial narratives she examines, they can be equally convincing if applied to some of the expanded narratives—material and linguistic, fictional and nonfictional, postcolonial or not—of a non-English-speaking Mediterranean country crucially positioned between the Global North and the Global South, which had, and continues to have, its own encounters with postcoloniality.[6] In short, the multifaceted selection of Italian-produced, environmentally revealing material-discursive stories and/or "storyworlds" (James 2015) I intend to engage with does not pertain exclusively to Italian studies but promises to offer cross-cultural environmental insights as illuminating as any other. And similarly, these narratives are able to play a comparably relevant, correlative role in influencing more sustainable worldviews and "enrich[ing] ecocritical discourse" (James 2015, 10)—not to mention thinking of Italian studies as not simply one of the provinces of modern languages.

Taking further inspiration from the still valid advice of Cheryll Glotfelty to aspiring ecocritical scholars to keep "one foot in literature and the other on land" (1996, xix), and complementing Iovino's groundbreaking ecomaterialist readings of Italy as a text and "'storied matter' endowed with narrative power" (2016a, 9), in the pages that follow, I thus draw an additional, alternative aesthetic and topographic map of the *bel paese*. Touching different locations on this map, I investigate a number of imaginative and physical terrains in modern and contemporary Italy where the places, bodies, and substances that have marked the nation's path toward modernity come to "meet" and interact. In particular, while striving to give equal weight to close readings (of fiction, nonfiction, visual works) and physical sites, all considered as texts rich with signs to be interpreted, I investigate the narrative eloquence and expressive energy of materials such as sulfur, petroleum, marble, concrete, asphalt, steel, and asbestos.

Building on the increasingly accepted idea of a decentralized agency that is shared among human and nonhuman entities, I reaffirm that these physical substances are not merely acted upon as inert stuff but rather crucially intertwined with human lives, corporalities, practices, meanings, and imaginative stories. As such, and as these materials interact and cooperate over time and space with human beings in shaping reality and effecting changes on the environment—just think of the material and discursive consequences of any ecocatastrophe—I argue that they necessarily contribute to making and, simultaneously, unmaking the country that is Italy today, affecting its socioenvironmental health in multiple ways.

As the still relatively few but crucial contributions in the field demonstrate, the Italian-Mediterranean context offers a promising and still largely unexplored object for ecocritical analysis, especially if viewed as an essential component of the larger dynamics of the Global South. Important books such as Marco Armiero and Marcus Hall's *Nature and History in Modern Italy* (2010), Monica Seger's *Landscapes in Between: Environmental Change in Modern Italian Literature and Film* (2015), Pasquale Verdicchio's *Ecocritical Approaches to Italian Culture and Literature* (2016), Iovino's *Ecocriticism and Italy* (2016), and *Italy and the Environmental Humanities* (2018), which I coedited with Iovino and Elena Past, have already stressed the fundamental point that in the past century, both the Italian landscape and the relationship of its inhabitants to this landscape have undergone rapid changes as a result of industrial, agricultural, and technological innovation. This uneven and disruptive shift from a mostly rural society to a modern one and its implicit effects on the environment began slowly in the early twentieth century, then exploded in the years of reconstruction and economic boom following World War II (mid-1950s to early 1960s). The several industrial accidents that have dotted the Italian territory since then—most memorably, the ones in Seveso (1976) and the incident at the Farmoplant plant in Massa Carrara (1988)—can only remind us that there have been actual explosions next to metaphorical ones.[7]

Needless to say, this multifaceted process of disruption is still underway, and in various guises, it continues to replicate itself in other developing parts of the world. Environmental historian Piero Bevilacqua observes that some of the agents involved in this process of disruption and subsequent socioenvironmental degradation were (and still are) human, but some were nonhuman, such as the "sulfur mines in Sicily and iron mines on the island of Elba," or "lead and zinc . . . and coal" in Sardinia, or the "steel and iron industry . . . in Terni . . . Piombino, Savona and the Bagnoli neighborhood in Naples," or the chemicals in the "triangle between Milan, Turin and Genoa" (2010, 21). In light of Bevilacqua's considerations, it is particularly appropriate, if not actually urgent, to interpret the expressive capacity of these agentic materials and, in general, matter's narrative potential in its eloquent interplay with human culture and discourse in the construction of meaning.

Although not specifically in relation to an Italian-Mediterranean context and not informed by new materialist theories, this sense of urgency may be traced back to Patricia Yaeger's initial remarks in a thematic issue of *PMLA* entitled "Literature in the Ages of Wood, Tallow, Coal, Whale Oil, Gasoline, Atomic Power, and Other Energy Sources" (2011), which helped me better define the topic of my research. As she and the other contributors focus on the roles and representation of such materials in a variety of literary texts and reflect on "the ways thinking about energy sources might transform our notions of literary periods" (Yaeger 2011, 310), they recognize the hermeneutical and methodological benefits of "addressing matter" and, by extension, of contemplating what Levi calls "the transversal bonds which link the world of nature to that of culture" (1989, 10; i legami trasversali che collegano il mondo della natura con quello della cultura).

More recently, such urgency is even more clearly articulated and underscored by Stacy Alaimo, who observes that in the current period of the

Anthropocene, it is impossible to avoid concerning ourselves with matter from a cultural theory perspective "when mass extinction, climate change, nuclear technology, industrial agriculture, urban and suburban sprawl, biotechnology, and the production of xenobiotic chemicals have profoundly altered the biophysical world" (2017, 13). Additionally, as she points out, chemicals, technologies, and various products transcorporeally traverse our bodies, literally and concretely making substances be in and of us. From its necessarily limited geographical perspective and focus, my book therefore represents one more attempt to address and respond to this kind of timely concern.

As should be at least partially evident by the names of the scholars mentioned thus far, my methodological approach is grounded in and inspired by recent theoretical paradigms and disciplines gathered under the large umbrella of the environmental humanities, from environmental history, social justice, and geocriticism to cultural/political ecology, econarratology, and posthumanism. In particular, I draw from the insights of the recent "material turn" within ecocriticism, a conceptual framework rooted in feminist theory and practice. Defined by Iovino and Oppermann as an approach that "examines matter both in texts and as a text, trying to shed light on the way bodily natures and discursive forces express their interaction whether in representations or in their concrete reality" (2014, 2), material ecocriticism frequently overlaps and "intersects with and has substantial affinities to [most of] the paradigms" mentioned previously (Zapf 2016, 72). Its most distinguished and, for my purposes, fertile features are its assignment of agency to nonhuman nature and its "emphasis on the narrative dimension of material reality" (Oppermann 2017, 290). By extending the notion of text to include all material formations, it expands ecocriticism's practical applications beyond nature writing, questioning traditional anthropocentric stances and theories by foregrounding that human beings are always embedded and embodied in their physical surroundings in a reciprocal relationship with agentic matter.

The philosophical genealogy that helped shape new materialisms is quite large and illustrious, and therefore, given the presence of excellent introductions to this conceptual field and its rich interdisciplinary cross-fertilizations, my overview here is very general and far from exhaustive.[8] Fundamentally, the new materialist theories that inform my ecomaterialist approach emerge out of and build on earlier feminist engagements with materialism. Informed by a continental, Deleuzian background, these theoretical positions question some of the perceived limitations and excesses of the linguistic turn and social constructionism and establish productive dialogues between cultural theory and the natural sciences. In addition to stressing matter's coproductive role in affecting human life and the sociocultural sphere, new materialisms also distinguish themselves for their more pronounced ethical component and fruitful links to posthumanism. Besides the scholars already mentioned, one should at least allude to Manuel DeLanda's excursions into "nonlinearity" and his ideas on the reciprocal interconnections between natural and historical processes, Bruno Latour's questioning of culture-nature and language-world dichotomies, Karen Barad's notions of "agential realism" and "intra-action," Jane Bennett's idea of a "vibrant materiality," Donna Haraway's material-semiotic view of the world, and Rosi Braidotti's enthusiastic posthuman scholarship. Although readers will find additional references to some of these thinkers in the following chapters, I wish to emphasize here the most relevant ethical-environmental implications of a scholarship that, despite its differences, nonetheless unanimously believes in repositioning the human in the more-than-human world and exploring the ramifications of this meaningful encounter.

In its invitation to view anthropocentrism—the tendency to assign agency and intentionality exclusively to human beings—as the intellectual force responsible for the devaluation of animals, plants, matter, and things, this new materialist turn aims to more ecologically and sustainably relocate the human horizontally and nonhierarchically. Namely, it repositions the human as just one more entity in a world that is constantly reconfigured by the "intra-action of discursive practices and material phenomena" (Barad 2007, 152)—in the flow of and in relation to other things, substances, and bodies. Thus this approach has far-reaching ethical consequences that go beyond and accompany any single aesthetic interpretation.

Given the context, it seems worth noting that Italian culture itself has not been alien to theoretical and political interrogations of the Western anthropocentric tradition that have a similar, implicit ecological flavor. Although they are from different disciplinary perspectives and backgrounds, especially over the past two decades or so, Italy has experienced a growth of innovative philosophical ideas that have helped us reflect on and reassess the intellectual legacy of Italian Humanism and the Renaissance.

The so-called New Italian Thought, through its own critique and questioning of the contemporary social and political world and its elaborations of the biopolitical paradigm, has certainly played a role in problematizing the humanistic dream of "man as measure of the world." In addition to the contribution of

Giorgio Agamben, already a familiar presence in Anglo-American academia, I am thinking in particular of the work of Roberto Esposito—specifically, his definition of Italian philosophy as "living thought" (pensiero vivente) and his individuation of a geophilosophy that, in his view, from the early sixteenth century materialistically opted "for an entanglement with the external world." As he puts it, "At the center of Italian philosophy, there is not the individual but the common world [mondo comune] in its inexhaustible vital power" (2010, 13).

Without pushing this too far or making superficial connections, I limit myself to observing that there are some points of potential synergy between Esposito's position and ecomaterialist thought. If it is true that among the latter's objectives is to reconsider ontologies by going outside and "meeting the universe halfway" (Barad 2007), it is also true that Esposito's "living thought" aims at getting involved with and is projected into the "world of historic and political life" (Esposito 2010, 12). Echoing some of the tenets of the new materialist scholarship, his affirmative biopolitics wishes to move beyond the limitations of the individual, humanistic being ("persona") in order to establish relations through different levels of reality and forms of life (human, animal, vegetal, material). Furthermore, it too implicitly questions exclusively anthropocentric worldviews and "welcome[s] life in all its different guises" (12).

Together with Esposito, although not associated with the "New Italian Thought," I approach the end of this introductory chapter by mentioning ethologist, zoo-anthropologist, and philosopher Roberto Marchesini. Starting with his *Posthuman: Verso nuovi modelli di esistenza* (Posthuman: Toward new models of existence; 2002) and, most recently, his *Alterità: L'identità come relazione* (Alterity: Identity as relation; 2016), Marchesini has aimed at overcoming the perception of human beings as isolated, self-referred entities impermeable to external contaminations. Rather, he pushes forth the notion of a heteroreferred, nonanthropocentric humanism that is not necessarily antihumanistic.[9] As such, some of his reflections will prove particularly useful and inspiring in the following pages.

Mindful of this eclectic theoretical background and, more tangibly, informed by the fundamental idea that we humans are immersed in the environment as much as the environment is in us, *Elemental Narratives* thus focuses on the Italian-Mediterranean natural-cultural dimension in order to draw attention to the intersection of human stories with those of the more-than-human worlds around and within us; it discusses some of the environmental implications of such an entangled situation. By taking an imaginary trip around the peninsula, it examines relevant "material narratives" emerging collectively from aesthetic forms (literary texts, artworks, documentaries), places (factories, construction sites, suburban areas), bodies, and materials. All these narratives insert the human in a productive dialogue with the nonhuman, increasing our understanding of "glocal" environmental challenges and, hopefully, helping us reassess some of our priorities for the future.

Sharing Hubert Zapf's belief that an "attention to the transformative role of the aesthetic in reshaping the ecocultural imaginary . . . is one of the crucial tasks of future ecocriticism" (2016, 50), I reflect on the combined role these localized material and cultural narratives—and storytelling in general—can play in raising awareness, changing perspectives, and shaping ideas about our engagement with the places we inhabit in the era of the Anthropocene and thus ultimately advancing the ecologization of our thinking and existence.

My first chapter, "Modernist Matters," begins with some theoretical reflections about situating a few Italian modernist authors within the current ecocritical (and specifically, new materialist) interpretative horizon. The interest of early twentieth-century modernism and avant-garde in matter, objects, and things is not critically new, nor is its reconsideration of the subject-object dualism, fascination with urban spaces, and expanded understanding of corporeality.[10] Less studied, however, is what such an interest may signify in ecocritical terms and in light of recent theoretical developments within the environmental humanities.[11] After arguing in favor of a specific elemental approach to modernism, I provide two sample case studies, briefly focusing on Scipio Slataper's autobiographical novel, *Il mio Carso* (My Karst; [1912] 1988), which is mostly set in the rocky northeastern mountains of the Friuli region, and on some of Luigi Pirandello's Sicilian sulfur narratives. Centered on F. T. Marinetti, I then invite further reflection on the general notion of matter, which is notoriously central to the futurist avant-garde. I draw attention to potential affinities, parallelisms, and, to use Deleuze and Guattari's term, "adjacencies" (1986, 7–8) between the way futurists—as they wrote about landscapes around Lake Garda, wood furniture, construction materials in Venice, and lava flows from Etna in Sicily—imagined the interrelated notions of nature, matter, and corporeality and some of the current positions of postmodern, posthuman material ecocriticism.

In the second chapter, "Slick Territories: Petroculture, Italian Style," I explore Italy's discursive and material encounter with petroleum. Initially, I return briefly to the early modernist decades of the twentieth century and to the futurist context, when it began its rise as a global source of energy. I first discuss some of the revealing "petro-texts" in *Il Gatto Selvatico*, one of the propaganda publications of Ente Nazionale Idrocarburi (ENI; Italian Hydrocarbon Corporation), and conclude by addressing material and discursive contaminations caused by oil in a novel and a recent documentary both set in Sarroch, in the south of Sardinia.

In chapter 3, "Apua Ma(t)ter: Narratives of Marble," I focus on the storytelling and agentic powers of a quintessential Italian material, Carrara marble, and the landscape of the Apuan Alps between the regions of Tuscany and Liguria, where the marble is found. The attempt to geopoetically connect the Apuan topography to the local mindscape allows me to discuss a number of fictional and nonfictional texts by indigenous authors from 1905 to 2015.[12] Like Michelangelo's *Pietà*, these texts literally emerge from this mineral, and at the same time, they become sites of resistance to politics of extraction and industrial exploitation.

The fourth chapter, "Steel and Asbestos: Stories of Toxic Lands and Bodies in Tuscany and Beyond," investigates the vibrancy of two closely connected substances and subverts stereotypical, pastoral, touristic depictions of Tuscany. My interest here is focused on how some recent fictional and nonfictional narratives have responded to long-standing yet often conveniently silenced eco-socio-biological crises in this region. I first consider the depiction of the steel town of Piombino as a site where environmentally poisonous entanglements between human actors and nonhuman matter take place, a reality represented in Silvia Avallone's novel *Swimming to Elba* (2012). Then I address an Italian version of what Alaimo calls "material memoir" (2010, 85), Alberto Prunetti's novel *Amianto: Una storia operaia* (Asbestos: A blue-collar story; 2012). Finally, departing from Tuscany to reach Casale Monferrato (in Piedmont), the "ur-site" of asbestos contamination in Italy and Europe, I consider other transmedial narrative dimensions and artistic performances (from journalistic blogs to "word theater") that chronicle the deadly consequences of hazardous industrial practices and asbestos exposure and manage to achieve some form of socioenvironmental justice.

Taking the site and material text of the Expo 2015 in Milan as the utopian/dystopian entry point for a final North-South journey along the peninsula, my fifth and final chapter, "Concrete and Asphalt: Geographies of Environmental Disruption in Modern Italy," begins by discussing the damaged ecologies and "widespread peripheries" of an extended Italian northeast in works by contemporary authors Giorgio Falco and Wu Ming 2 (alias of Giovanni Cattabriga).

Partially following the narrative walking route Wu Ming 2 carved through the Po Valley in his *Il sentiero luminoso* (The shining path; 2016), I investigate how cement and building practices in general are central issues in *Violazione* (Violation; 2012), a novel by Alessandra Sarchi set in the region of Emilia Romagna. Sarchi's novel treats the violence and danger naturally present in but also cynically layered onto the land, our twisted relationship with the nonhuman other, and in short, the essentially unethical dimension of the space we inhabit. Issues of environmental justice, soil consumption, and abusiveness recur in the final part of this chapter. Here I establish an imaginary link between Tuscany and the southern region of Calabria, attempting a joint consideration of Simona Baldanzi's novels and her inquiries into the socioenvironmental impact of new tunnels and Treno Alta Velocità (TAV; a high-speed train) infrastructures in Tuscany's Mugello area as well as Mauro F. Minervino's poetic reflections on Calabria in his compelling "road narrative" *Statale 18* (State road 18; 2010).

In a frequently cited remark, Robert Kern writes that "ecocriticism becomes most interesting and useful . . . when it aims to recover the environmental character or orientation of works whose conscious or foregrounded interests lie elsewhere" (2000, 11). My research shares his perspective, and as the chapter descriptions should suggest, with very few exceptions, it does not address narratives that have a specific environmental agenda.

Along a similar line, and in spite of the potential appropriateness of some of their works, my discussion does not address (either directly or tangentially) a number of canonical authors such as Italo Calvino, Pier Paolo Pasolini, or Gianni Celati, who have already been the object of ecocritical attention (see works by Seger, Iovino, Luisetti, and Barron). This explains why, for example, readers will not find a classic like Calvino's short novel *La speculazione edilizia* (A plunge into real estate; 1991b) or a discussion of Celati's take on the Po Valley in my "Concrete and Asphalt" chapter. Likewise, in the chapter on oil, I barely skim the surface of Pier Paolo Pasolini's *Petrolio* (Petroleum; [1972] 1992). While these names remain crucial in Italian ecocritical discourse, *Elemental Narratives* implicitly demonstrates that "storied" environmental concerns are more pervasive in Italy than previously thought by introducing lesser-known and unexpected figures to such a discourse. As such, it boosts the relevance and visibility of Italian studies in the larger field of the environmental humanities.

"This little trick with amber was a sign to decipher. . . . It will never be possible to eliminate all risk nor to solve all problems, but every solved problem is a victory in terms of saved human lives, health, and wealth" (Levi 1989,

138; Il giochetto dell'ambra era un segno da decifrare. . . . Non si riuscirà mai ad annullare tutti i rischi né a risolvere tutti i problemi, ma ogni problema risolto è una vittoria, in termini di vite umane, salute e ricchezze salvate), writes Primo Levi in his collection *Other People's Trades* (1989). Interestingly, even though Levi refers only to human lives, I am quite confident that given his obsession with animals and matter, the addition of *nonhuman* would not be unwelcome here. This said, what amber eventually reveals is electricity: "a force that would change the face of the world" (138; una forza che avrebbe mutato la faccia del mondo) but that, if not respected, also contains a harmful potential.

To a certain extent, and despite their differences, all the matters discussed in the following pages are a bit like amber. Depending on how we interact (or, better, "intra-act," to quote Barad 2007, 33) with them, they can solve problems or create disruptions and trouble for human and nonhuman lives, health, and wealth. Continuing to decipher the signs they send and to listen and learn from their emerging stories may then be a good, humble strategy after all. If it does not immediately save lives, then it may at least raise awareness, keep humankind in balance with the environment, ameliorate attitudes, and prevent some of the calamities that we (some of us more than others) excel in bringing upon ourselves and our planet. After all, on this planet, as Levi wisely sensed, "everything seems stable and is not . . . [and] awesome energies . . . sleep a light sleep" (1989, 110; tutto sembra stabile e non è . . . [e] spaventose energie . . . dormono di un sonno leggero).

1
Modernist Matters

Sulfur and Rocks: Thinking Toward an Elemental Modernism

For a book that wishes to approach Italy and modern Italian culture through an ecocritical lens and, more specifically, focus on some of the materials that marked the country's postunification path toward modernity, it seems fitting to begin with a chapter on Italian modernism.[1] Indeed, the modernist period is a natural object of exploration, given its internal contradictions and tensions, its general sense of crisis, and its conventional associations with technological progress, nationalism, human mastering of nature, the change of relationship between city and countryside, feelings of alienation, and the objectification of individuals.

Additionally, as Mimi Sheller notes in her fascinating *Aluminum Dreams*, the making of twentieth-century modernity is closely entwined with the sociocultural history of the materials that helped shape it and define it: aluminum, of course, but also "iron, steel, plastics, glass and cement" (2014, 3). The quote by Arnaldo Mussolini (brother of Benito) that opens the book perfectly bolsters her argument, capturing the centrality of materials that are at the same time objects but also subjects and agents of modernity: "We have often said: just like the nineteenth century was the century of iron, heavy metals, and carbon, so the twentieth century should be the century of light metals, electricity, and petroleum" (quoted in Sheller 2014, 1). Most significantly for my objectives, Sheller is also quick to point out the "conundrum" aluminum poses. It is "so full of promise as a technological solution across a range of applications yet has caused so much unacknowledged environmental and human harm" (4).

From a more theoretical perspective, focusing on modernism also allows us to reassess the more immediate, mostly present-oriented, and intuitively recognizable contiguity between ecological crisis and postmodern sensibility, implicitly expanding in temporal and spatial terms the range of authors, texts, and places considered for analysis. After all, as Hubert Zapf observes, "Modernism and postmodernism ... can no longer be seen as dichotomies, but as interconnected ways of making sense of a constantly changing world, a world which always transcends the categories that are applied to it" (2010b, xii). There are moments in which modernist thought and aesthetic practices become less sharply distinguishable from their postmodern counterparts as they engage with nonhuman alterity, question anthropocentric viewpoints, and ultimately delineate an "ethics that is embedded in material relations between human and nonhuman, culture and nature, and that gains its force through experiments in both content and form" (Ryan 2015, 300). The individuation of some of these moments as they pertain to the Italian context is certainly among this chapter's and, more generally, this project's main objectives.

Significantly, from different critical angles, even traditional (here in the sense of not ecocritically oriented) literary scholarship that focuses on European and Anglo-American modernism repeatedly observes how modernist artists are captivated by the physical world of natural or technological "objects, matter and things," to quote from the subtitle of Bill Brown's essay on the materialities of modernism. Expectedly, this kind of scholarship usually underlines the object's and the material world's aesthetic roles and "ontological amplitude" (2013, 289). Namely, it stresses the material world's epistemological, revelatory, existential relevance in relation to the production of human subjectivity and does not concern itself with specific environmental issues. And yet I find it revealing how some scholars of modernism already hint—more or less intentionally and explicitly—at some of the larger ecological and ecocritical implications of the modernist tendency toward "thinking matter" that this chapter aims at addressing in its Italian, mostly futurist declension.

At the beginning of his classic *All That Is Solid Melts into Air*, for example, Marshall Berman notoriously defines modernism as "any attempt by modern men and women to become subjects as well as objects of modernization, to get a grip on the modern world and make themselves at home in it" (1982, 5). As I choose to view this remark here, Berman might be questioning the dividing line between people and things, human and nonhuman, while also suggesting that the difficulties and challenges involved in any modernization process may leave someone homeless. His reference to the dwelling (oikos) that human beings try (and sometimes fail) to make for themselves in the modern world has a special resonance if viewed from an ecocritical perspective.

As they implicitly allude to tangible processes of spatial control and dominance, Berman's words invite us to think less abstractly about the relationship between modernity and modernism and, in turn, to focus on the incrementally complex intertwining of things, substances, and people in these decades. In other words, I take Berman's definition of modernism literally as an invitation to pay attention to the textual, spatial, and sociocultural (material-discursive) spaces that are constantly created, affected, or torn apart by the conflictual "maelstrom" of the modern (1982, 15), either in its characteristic urban "landscape[s] of steam engines, automatic factories, railroads, vast industrial zones; of teeming cities that have grown overnight, often with dreadful human consequences," or in its alternative, nostalgic "returns to nature" (18–19).

A similar invitation can ostensibly be found in Douglas Mao's *Solid Objects: Modernism and the Test of Production*, a book that explores Anglo-American modernism's fascination with the world of objects and its engagement with materiality and whose central argument could lend itself to ecocritical expansions. In addition to Mao's definition of Ezra Pound as "ecologically minded," one finds inspiration in his observation that for modernists, the object world is simultaneously immune "to thinking and knowing . . . out of reach of human persuasion" but also "not . . . out of reach of the human power to destroy" (1998, 9). Similarly thought provoking is his realization that literary modernism often conveys an ambivalent and anxious attitude toward the notion of production, seen as a meaningful way of "leaving a material trace on the object world . . . of valuing human accomplishment"—but one that is full of "liabilities" (11), "since any kind of making, be it handicraft or mass production, could also be understood as a violation of the nonhuman world by an increasingly predatory and imperialistic subjectivity" (Princeton University Press, n.d.). Without oversimplifying Mao's argument, these statements already tell us something about the difficulty of separating issues dealing with modernist relations between humans and objects (in the most comprehensive sense of the word) from environmental questions.

More recently, Eric Falci has used the convenient expression "elemental modernism" to synthesize both the difficulty and the complexity of these relations. In particular, he uses it to describe both a mode that is "central to modernist poetry" and also a "mid-century tactic . . . not simply a respinning of modernist primitivism . . . in which poets chart the complex difficulties and horrors of modernity by way of a broken return to elemental forces—to the environment and the animal world, to the natural history of earth, and to early civilizations, myths, and religions" (2015, 65).[2]

In the Italian context, the category of modernism did not find a particularly welcoming environment. For starters, this was a country where the signs of modernity and the effects of modernization arrived later than other European nations and the United States. Moreover, until recent critical debates and reassessments, Italian intellectuals and literary critics peculiarly avoided the term, adopting instead a number of other "isms" and definitions (decadentism, crepuscularism, futurism, avant-garde) to indicate the several, transnational, heterogeneous projects of cultural regeneration that roughly characterize the period going from the end of the nineteenth century to the 1940s. In any case, just to provide a simple working definition, we may state with Raffaele Donnarumma that in Italy and elsewhere, "modernism becomes the name of the twentieth-century canon up until the Second World War" (2012, 13) and that, in terms of literary forms, ideologies, and stylistic features, it is equally (and often contradictorily) informed by both continuity and rupture.

So despite the fact that "in Italian literature the category of 'modernism' has never been really at home" (Valesio 2004, ix), there is no doubt that several artists in Italy—like those in England, France, or the Americas—reacted to the dramatic existential and sociocultural changes brought by technological and protocapitalist modernity, and as such, they can be included in such an "elemental" mode. Within the larger modernist preoccupation with human and material embodiment, a number of Italian artists grappled with the limits of the body and its entanglements with specific substances and landscapes while also imagining a "return to elemental forces" (Falci 2015, 65).

Besides Marinetti and the futurists, whose engagement with the natural world and the general notion of matter will be addressed later in this chapter, and figures such as Enrico Pea, Carlo Emilio Gadda, and Massimo Bontempelli, who will be evoked more or less at length in the following ones, I briefly consider here two examples drawn from literature that show both how such elemental tendencies manifest themselves within Italian modernism and how an ecological reading may enrich their interpretation. These are Luigi Pirandello's short stories dealing with sulfur mining in Sicily and, moving toward the peninsula's northeast, the first part of Scipio Slataper's lyrical autobiography *Il mio Carso* (My Karst; [1912] 1988). On the one hand, stories such as Pirandello's "Il fumo" (The smoke; 1994d), "Ciàula scopre la luna" (Ciàula discovers the moon; 1994a), "Formalità" (Formality; 1994b), and "Fuoco alla paglia" (Fire to the straw; 1994c) convey multifaceted accounts of the dynamics and damages that sulfur extraction, processing, and commercialization brought to Sicilian landscapes, individuals, communities, and economies in the nineteenth and early twentieth century, which left them all as literal and metaphorical wastelands.

On the other hand, Slataper's novel assigns a powerful cognitive role to the rocks and the mountainous landscape that provide a part of its background that is essential to the eventual maturation of the protagonist. Significantly, the text begins by affirming the main character's urge to immerse himself in and be energized by nature in the Karstic plateau surrounding Trieste before he moves to the city and attempts to "poetically" rejuvenate what he perceives as a sick and decadent metropolis, "weakened by its excessively commercial character, choked by the fumes of industry, and scarred by the uniformity of countless grey buildings" (Bond 2016, 7).[3] His intimate interpenetration with the natural world is vividly described in the following passage:

> The earth on which I slept during my deep nights knew me. . . . I ran with the wind. . . . I plunged head-first into the river to quench my skin's thirst, drench my throat, nostrils and eyes with water and, as I swam under water with a pike-like open mouth, I swallowed large gulps. . . . As I walked, I looked at everything with a brotherly love. . . . Mount Kal is a stony quarry but I feel fine with him. My overcoat sticks to the rocks like meat on ember . . . my hands sink into his edges which wish to conjoin with my bones.
>
> Mi conosceva la terra su cui dormivo le mie notti profonde. . . . Correvo col vento . . . mi buttavo a capofitto nel fiume per dissetarmi la pelle, inzupparmi d'acqua la gola, le narici, gli occhi, e m'ingorgavo di sorsate enormi, notando sott'acqua a bocca spalancata come un luccio. . . . Camminando guardavo tutto con affetto fraterno. . . . Il monte Kal è una pietraia. Ma io sto bene con lui. Il mio cappotto aderisce sui sassi come carne su bragia . . . le mie mani s'incavano contro i suoi spigoli che vogliono congiungersi con le mie ossa. (Slataper [1912] 1988, 63, 64, 68)

This description of the protagonist's fusion with the Karstic environment, in which all the elements (water, earth, air, fire) are evoked ahead of his descent to the city as "Pennadoro"[4]—before his subsequent entrance into an exclusive dimension of cultural order and engagement with Trieste's intellectual life—may be considered in light of some of Jeffrey Cohen and Lowell Duckert's inspiring statements in their *Elemental Ecocriticism*. As they write in their introduction, "Elemental matter is inherently creative, motile, experimental. . . . Elements-as-fundamentals trigger beginnings, catalyze arrangements that resist totality. . . . The elements are hostile. . . . Except, of course, the house is composed of domesticated earth, air, fire, water.... They are the outside that is already within, the very stuff of cosmos, home, body, and story" (2015, 3, 9, 11, 13).

These remarks suggest that Karst and Trieste, the natural and cultural homes and dimensions that characterize, respectively, the initial and conclusive parts of the novel, may not be irreconcilable opposites but rather mutually constitutive. A recent article by Emma Bond, in which she argues that the protagonist's drinking and merging with "the repugnant physicality of his fellow drinkers" in a tavern in Trieste allows him to "accept a more fluid, multifaceted view of the external world as well as of the self" (2016, 9, 16), implicitly suggests that any growth in him, as well as any new way of knowing, depends on his proximity to a world that is different from his own and on his engagement with alterity.

But it may not only be the protagonist's adult immersion into an urban "landscape of intoxicated geography" that causes him to eventually abandon his limited, polarized self-perception ("between a high spiritual or poetic side and a low animal or instinctual side") and "embrace and accept a more complex view of the fragmented nature of both the self and of society as a whole" (Bond 2016, 3). I contend that his youthful, corporeal entanglement and elemental encounter with the Karstic natural landscape is similarly relevant for altering his disposition and providing him with a clearer, though still "in process," knowledge of his physical surroundings. Namely, the moment when he quenches "his skin's thirst" with water from a stream rather than with alcohol triggers his onto-epistemological beginning. It is also such a primitive, elemental plunge that eventually opens for him a new intellectual and existential path, allowing him to "love and work" (Slataper [1912] 1988, 153; amare e lavorare) in the modern city and to complete his transformation into a poet who attempts to voice the difficulties and contradictions of modernity as both *kaos* and *kosmos*.[5]

It is worth mentioning that a recent interpretation of Slataper's novel, which I learned about only after I had drafted this book, further supports my embryonic reflections and the kind of approach outlined here. Departing from similar theoretical premises, even though it eventually provides a much more in-depth and nuanced reading of the whole novel rather than a general hermeneutical key, Deborah Amberson's essay "Temporalities of Stone in Scipio Slataper's *Il mio Carso*" builds on Aldo Leopold's expression "thinking like a mountain" and Jeffrey Cohen's lithic ontology in order to emphasize Karst's role in shaping the protagonist's growth and sense of awareness, leading to his ultimate "comprehension of the mutual entanglement of all beings" (Amberson 2018, 2).

I wish to underline here Amberson's convincing observations on the vitality of Karst, viewed as "a material space that blurs potential divides between the organic and the inorganic, between land and sea, between barrenness and fecundity, and, finally between past and present" (3). Amberson also assigns Karst a crucial role in forging "a new geological morality" through which the protagonist learns that he is a "man among men" (uomo tra gli uomini) but also part of a "stone infused humanity . . . 'extracted from the earth' where humans had slept as the 'dark hard heart of the earth'" (12; buio cuore duro della terra).

A parallel tension and overlapping between self and other, order and disorder, unity and fragmentation, *kaos* and *kosmos* also inform Pirandello's work as he articulates his own vision of modernity. Set in the desolated, arid countryside around Agrigento, in an area historically central to the sulfur-based economy between the middle of the nineteenth and the first decades of the twentieth century, his short story "Il fumo" stands out as particularly compelling in the elemental context of this chapter. The title refers both to the material by-product of intensive burning practices necessary to obtain sulfur and to the metaphorically clouded vision of those who see these fires and yet are unable to avoid their deleterious consequences. This story also shows a prescient understanding of the dangers and implications of early capitalist industrial practices, and it prefigures current environmental concerns about air quality.[6]

I approach this text in light of Andreas Malm's insights in a recent essay in which—refusing to get completely rid of "close reading [and] close writing in a warming world" (2017, 127)—he links together climate change, fossil fuel fiction, and Walter Benjamin's theory of dialectical images. As Malm puts it, quoting Benjamin, "A work or even a fragment of fossil fuel fiction might be read as an object—more precisely a monad—'into which all the forces and interest of history enter on a reduced scale.' It might be retrieved as a *dialectical image* whose import can be fully recognized only now that we stand on the verge of uncontrollable climate change" (128). This strategy, Benjamin maintains, allows us to learn about the past and, at the same time, reveals something crucial about our present and "bring[s it] into a critical state" (quoted in Malm 2017, 135).

Now, even though traces of sulfur can be found in both coal and oil, sulfur is not usually considered a fossil fuel in itself.[7] However, the distance that separates sulfur from petrol and coal diminishes when one considers the fact that this substance, similar to coal, was extracted from mines by the poorest and most oppressed workers (*i carusi*); that, like oil, it was then used to literally propel agriculture and a number of factories forward for several decades in the eighteenth and early nineteenth centuries (producing gunpowder, textiles, matches, and pharmaceuticals); that it has been mythically associated with scorching infernal depths; and that it is highly flammable and toxic when burnt.[8]

Keeping this in mind, a number of fragments in Pirandello's novella, especially in its opening and concluding sections, may reveal "dialectical images" in Benjamin's sense. Namely, they shed light not only on the localized and time-specific environmental impact of a specific fossil-fuel-like substance and industrial model but also on subsequent geographically disparate and yet similar exploitative forms of industrialization and commodification and their compounded planetary effects. The story's powerful initial passages—which focus on the workers' emergence from the holes of the sulfur quarries that punctuate a hot, physically oppressive, asphyxiating landscape—provide an apt case in point:

> As soon as the sulfur miners came up panting and with their bones aching from the bottom of the "hole," they immediately searched with their eyes for the green over there, in the distant hill that closed the wide valley on the west. Here instead, the arid mountainsides, livid with scorched tuffs, no longer had a grass blade, they were pierced by anthill-like sulfur quarries and all of them were burnt by the *smoke*. . . . The sight of all that green in the distance also eased the pain of breathing, the sour oppression of *smoke* clinging in one's throat, which caused the cruelest spasms and wheezes of asphyxia. . . . The *carusi* . . . thought of a life in the countryside, content, without risks, without serious hardships there in the open, under the sun, and envied the farmers. . . . Underground, they looked like many busy dead people. . . . Here was their enemy: the devastating *smoke*. . . . The countryside was there, lying in the sun, for everyone to see . . . the sulfur mine, instead, was blind and woe betide slipping inside it.

> Appena i zolfatari venivan su dal fondo della "buca" col fiato ai denti e le ossa rotte dalla fatica, la prima cosa che cercavano con gli occhi era quel verde là della collina lontana, che chiudeva a ponente l'ampia vallata. Qua, le coste aride, livide di tufi arsicci, non avevano più da tempo un filo d'erba, sforacchiate dalle zolfare come da tanti formicaj e bruciate tutte dal *fumo*. . . . La vista di tutto quel verde lontano alleviava anche la pena del respiro, l'agra oppressura del *fumo* che s'aggrappava alla gola, fino a promuovere gli spasimi più crudeli e le rabbie dell'asfissia. . . . I *carusi* . . . pensavano alla vita di campagna, vita lieta per loro, senza rischi, senza gravi stenti là all'aperto, sotto il sole, e invidiavano i contadini. . . . Parevano, sottoterra, tanti morti affaccendati. . . . Era là il loro nemico: il *fumo* devastatore. . . . La campagna era lì, stesa al sole, che tutti

potevano vederla ... la zolfara, all'incontro, cieca, e guaj a scivolarci dentro. (Pirandello 1994d, 45–46)[9]

Rhetorically structured as a series of juxtapositions (below/above, surface/depth, near/close, darkness/light, sickness/health), this image of fatigued zombies working underground in a place that feels and smells like hell and who are eager to find an elusive, utopian relief in the not-yet but likely soon-to-become hell of the green countryside above resembles what Malm (referring to Ghassan Kanafani's novella *Men in the Sun*) calls "the fate of the poor in a warming world" (2017, 128). At the same time, the image hints at a continuity between corporeal and environmental degradation, bodies and landscapes.[10] The main ecocritical value of this story is precisely that, as the pervasive "material allegory" (129) of the smoke suggests, if human beings persist in what Mezzadra and Neilson (2017) would call blind extractivist practices—be they related either to the material removal of sulfur from the earth's depths or, in an extended sense, to the socioeconomic and psychological draining of the community at the center of this operation—the infernal darkness below will soon spread everywhere aboveground, eventually affecting everyone, regardless of social position or wealth.[11]

Part of this novella's narrative tension and resonance in our present anthropo- and capitalocenic time derives from the fact that its protagonist (Don Mattia Scala), similar to his fellow small landowners, is forced to negotiate with mining companies and governmental agencies while simultaneously striving to stage some resistance to their power. Conscious of the dangers of the "devastating *smoke*" (Pirandello 1994d, 46; *fumo* devastante),[12] he finds himself entangled in a transnational circle of greed, speculation, and profit initiated by and revolving around sulfur. In the end—desperate and under emotional and financial pressure—he can only betray his friends and moral principles and, making an already bad situation worse, burn down "the countryside that could no longer belong to him" (Pirandello 1994d, 77; la campagna che non poteva più esser sua). The following passage reveals the steps that are conducive to such a tragic conclusion:

> We work, we kill ourselves to excavate it, then we transport it down to the marinas, where many English, American, German, even Greek steamboats are ready to swallow it with their gaping holds; they throw us a nice whistle, and goodbye! ... And our wealth in the meantime, that which should be our wealth, goes away like this from the veins of our gutted mountains, and we are left here like many blind fools, with our bones

broken by fatigue and empty pockets. Our only gain: our countryside scorched by the smoke. . . . In twenty years, those who would come after him [Don Mattia Scala] . . . would see a bald, parched, bruised hill, pierced by the sulfur mines. . . . And he, he would be the ravager of the green hill!

Lavoriamo, ci ammazziamo a scavarlo, poi lo trasportiamo giù alle marine, dove tanti vapori inglesi, americani, tedeschi, francesi, perfino greci, stanno pronti con le stive aperte come tante bocche a ingojarselo; ci tirano una bella fischiata, e addio! . . . E la ricchezza nostra, intanto, quella che dovrebbe essere la ricchezza nostra, se ne va via così dalle vene delle nostre montagne sventrate, e noi rimaniamo qua, come tanti ciechi, come tanti allocchi, con le ossa rotte dalla fatica e le tasche vuote. Unico guadagno: le nostre campagne bruciate dal fumo. . . . Fra vent'anni, quelli che sarebbero venuti dopo di lui . . . avrebbero veduto un colle calvo, arsiccio, livido, sforacchiato dalle zolfare. . . . E lui, lui sarebbe stato il devastatore della verde collina! (54, 77, 78)

While there are substantial differences between early twentieth-century protocapitalist industrialization models and the contemporary forms of industrial capitalism discussed by Mezzadra and Neilson, certain fundamental dynamics of exploitation and dispossession as well as ideals of unlimited expansion are likely at the roots of both movements and historical moments. As a consequence, it is tempting to read these additional dialectical images of material excavation, socioenvironmental destruction, and scarred landscapes—together with Don Mattia's impotence and feeling of being exploited—in light of Mezzadra and Neilson's words when, reflecting on "the multiple frontiers of extraction" they observe, "It is not only when the operations of capital plunder the materiality of the earth and the biosphere, but also when they encounter and draw upon forms and practices of human cooperation and sociality that are external to them that we can say that extraction is at stake" (2017, 4). Although in an embryonic manner, I suggest that these kinds of operations and expanded forms of extraction are also present in Pirandello's "Il fumo." This text illustrates the correlation between the draining of natural resources from the earth and the withering of relationships of trust, cooperation, and sociality between human (and nonhuman) beings. In this sense, the dark mining holes encountered at the beginning are not only literally piercing a specific extension of mineral-rich land in Sicily. As one follows Don Mattia through the apparently independent but actually related events that precede his final decision to "become sulfur" and send his land up in smoke rather than lose to a greedy ex-business partner,[13] one realizes that throughout the narrative, other holes have been made in the sociocultural fabric of his community. To put it otherwise, holes have been "cut through patterns of human cooperation and social activity" (Mezzadra and Neilson 2017, 10).

In light of these considerations, "Il fumo" may ultimately be considered as a modernist "disaster narrative." It is one that "discloses the material and moral entanglement of human and nonhuman actors and factors in the etiology, unfolding, and aftermath of catastrophes" (Rigby 2014, 214) and that, along with Benjamin, brings out issues that are critical for our present at a time when the extraction practices it depicts at a small, local scale are becoming "increasingly invasive" (Mezzadra and Neilson 2017, 17), global, and always more challenging to confront.

Given this scenario, it is only fitting that the final apocalyptic image of the story mirrors its dismal opening, but the conclusion is narrated in the conditional tense and projected forward, somberly reminding every reader (past, present, and future) what is to come unless one learns to literally and materially "see through the smoke":

> [Don Mattia] . . . looked around at the trees . . . those centennial olive trees. . . . He imagined how all those leaves which now were alive would wrinkle at the first harsh vapors of the sulfur mine, opened there like a mouth of hell; then the leaves would fall; then the naked trees would turn black, then they would die, poisoned by the smoke of the furnaces.

> [Don Mattia] . . . guardò attorno gli alberi . . . quegli olivi centenarii. . . . Immaginò come tutte quelle foglie, ora vive, si sarebbero aggricciate ai primi fiati agri della zolfara, aperta lì come una bocca d'inferno; poi sarebbero cadute; poi gli alberi nudi si sarebbero anneriti, poi sarebbero morti, attossicati dal fumo dei forni. (Pirandello 1994d, 78)

At this point, while noticing how this image might also refer to one of the natural landscapes soon to be torn apart by World War I, we may also observe that around the same time in which Pirandello showed a substantial level of ecological awareness,[14] another Italian modernist writer, F. T. Marinetti, was apparently "seeing through the smoke" in a very different way. He was not only notoriously "glorifying war" (Marinetti 1990b, 11; glorificare la guerra) but also imagining mixing sulfur's explosive power with other substances to blow up traditions and reach some new sort of cultural enlightenment: "Whirlwinds of aggressive dust; blinding fusion of sulfur, potassium and silicates for the windows of the Ideal! . . . Fusion of a new solar globe that we'll soon see shining" (Marinetti 1990l, 15; Turbini di polvere aggressiva; acciecante fusione di zolfo, di potassa e di silicati per le vetrate dell'Ideale! . . . Fusione d'un nuovo globo solare che presto vedremo risplendere!).

In the following section, I will thus consider from an ecomaterialist perspective this figure who recommended that we "embrace the life of matter" (Marinetti 1990i, 48; abbraccia[mo] la vita della materia) and put "matter's lyrical obsession" (50; ossessione lirica della materia) in place of "human psychology" (50; psicologia dell'uomo).[15] A more detailed investigation of Marinetti's and futurism's relationships with petroleum—together with aluminum, one of the "lubricants" of modernity—will then continue in the second chapter, which is entirely dedicated to oil.

Ecofuturism? Nature, Matter, and Body in F. T. Marinetti

> We want to sing the man at the wheel, the ideal axis of which crosses the Earth, itself hurled along its orbit.
>
> Noi vogliamo inneggiare all'uomo che tiene il volante, la cui asta ideale attraversa la Terra, lanciata a corsa, essa pure, sul circuito della sua orbita.
>
> —F. T. Marinetti, "Fondazione e manifesto del futurismo"

> First of all, let's put an end to the boasted superiority of the human.
>
> Anzitutto finiamola colla vantata superiorità dell'umano.
>
> —F. T. Marinetti, "Il poema non umano dei tecnicismi"

Both quotes above accurately represent the frequently contradictory thought of F. T. Marinetti, the champion of Italian futurism, one of Europe's most influential avant-garde movements: the exaltation of speed and technological control over a submissive Earth mastered by "man" versus a rejection of the "boasted superiority of the human" (Marinetti 1990d, 1189). It seems that we could hardly be further away from an ecological horizon. But is this really the case?

Certainly, Marinetti's works do not display an environmental consciousness in the modern, common sense of the expression, and his relationship with the natural world is definitively more conflictual than sustainable. Futurism's *machinolatria* and vitalistic cult of violence, on the one hand, and its anxiety toward the power and forces of nature, on the other, translate in repeated textual attempts to transform the latter into an enemy, aggressively battle with it, tame it, and overcome it. Marinetti was a staunch nationalist and imperialist figure with a technophilic disposition who wished to urbanize and modernize natural landscapes rather than contemplate their beauty or appreciate their spiritual dimension. He desired to "glorify war" (1990b, 11; glorificare la guerra), "kill the moonshine" (1990l, 14; uccid[ere] il chiaro di luna), achieve the "new beauty [called] geometric and mechanic Splendor" (1990i, 98; nuova bellezza che io chiamo Splendore geometrico e meccanico), crush the dream of an "absurd return to the life in the wild" (1990e, 161; ritorno assurdo alla vita selvaggia), and even conquer the "Gruff Wind leader of the Forces of the Past" (1990k, 1017; Vento Burbero comandante delle Forze del Passato), which he perceived as an obstacle to overcome. As scholars have observed, Marinetti viewed nature as a resource to be conquered and civilized—a corpus to be mastered and exploited in order to affirm the bellicose virility of the masculine subject. In Poggi's words, nature for Marinetti is "the locus of the feminine and the maternal [that] must be opposed and displaced by both the machine, and its symbolic ally, matter (sheer dynamic physicality)" (1997, 24). Likewise, Marja Härmänmaa's assessment is that "in the clamor of the modern metropolis, the silent voice of Nature was forgotten.... If we want to learn something from the errors of the past, the case of 'Marinetti' is worth remembering" (2009, 358).[16] It is certainly difficult to disagree with her if, for example, one recalls the end of *Mafarka the Futurist*, when Gazurmah takes his superhuman flight with his masculine "breath" (Marinetti 1998a, 206; soffio) that "bends the forests" (206; piega le foreste) and his wings that reduce a feminized sky to a "long scarf continuously tied and untied" (206; lunga sciarpa continuamente annodata e snodata). Later, a similar passage in "L'aeropoema del Golfo della Spezia" (The aeropoem of the Gulf of La Spezia) describes an airplane copiloted by the writer as it lifts off and pierces a personified feminine atmosphere in an attempt to possess a

> *delirium of the atmosphere that slowly unlocks her smooth, turquoise, hard thighs so very very hard here she gives herself to the incalculable pangs opening wide of voluptuousness and irascible padding (which) I penetrate inside.*

deliirio dell'atmosfera che disserra a poco a poco le sue coscie liscie turchine dure tanto tanto dure ecco si dà tutta aperta verso lo spasimo incalcolabile spalancarsi di voluttà e ovatta iraconda vi penetro dentro. (Marinetti 1990f, 1097)

Nevertheless, a first step to reconsidering the interrelated notions of nature, matter, and body in Marinetti's work from a perspective that is receptive to current ecocritical concerns and theories might be to recall Anne Raine's remark that "the appreciation of and concern for the natural world may be more central to literary modernism than critics have recognized" (Raine 2014, 99). At the same time, it helps to keep in mind the "Janus-headed configuration of literary and artistic Modernism" and the "paradoxical dynamics" characterizing European culture at the turn of the twentieth century—that is, a culture able to generate often contradictory and antipodal responses "with respect to social and cultural norms" (Moroni 2011, 117–22). Thus it is especially important when, on rare occasions in Marinetti's texts, nature "becomes the hero (or heroine) that cruel, materialistic technology victimizes" (Blum 1996, 141). This interpretative path is taken by Cinzia Sartini Blum in *The Other Modernism*. Focusing on passages from *The Untameables* (1998b), *Spagna veloce e toro futurista* (Speedy Spain and a futurist bull; 1990k), *Il fascino dell'Egitto* (Egyptian fascination; 1990c), and "Il poema non umano dei tecnicismi" (The nonhuman poem of technicalities; 1990d), Blum notes that the natural realm—in the form of a desert oasis and a lake, a bull, palm trees, a cane thicket, and a submerged forest, respectively—is positively and nostalgically associated with the notions of peace, seduction, the maternal, "the transforming power [as well as the danger] of progress," heroism, compassion, and regeneration (1996, 139).

The list of texts displaying this transforming power could be further enriched. There are, for instance, a few revelatory excerpts from the collection *Novelle colle labbra tinte* (Stories with painted lips; 2003d). The protagonist and narrator in "Meandri di un rio nella foresta brasiliana" (Meanders of a stream in the Brazilian forest), for example, confesses not only a Pan-like desire to merge his body with the mixed organic matter of the surrounding forest and be thus reduced to an "instinctive state" (Marinetti 2003c; stato di istinto); he also assigns a vitalistic energy (of the kind futurism more frequently associates with machines and technology) to the flowing water of the river itself. The latter, in turn, is set in a symbolic opposition to the obstacle represented by "heavy" memories and regrets: "The horizontal effort of the water signifies also the tenacious energy of life while bulky and braking memories rain down" (72; Lo sforzo orizzontale dell'acqua significa anche l'energia tenace della vita mentre diluviano i ricordi ingombranti e frenatori). A similar wish to become one with the natural surroundings occurs in "11 baci a Rosa di Belgrado" (11 kisses to Rose of Belgrade; 2003a) and "La logica di Ahmed Bey" (The logic of Ahmed Bey; 2003b). The former paradoxically suggests both a loss and an extension, or a strengthening, of the lovers' human, anthropomorphic nature as they embrace and dissolve their bodies into the sea, thus incorporating it, ingesting it, and assuming the features of its inhabitants. These events take place while the surrounding rocks are in turn anthropomorphized and lose their inorganic nature:

> We are part of the powerful marine whole. . . . I drink a turbulent salvage of rhizopods. I have on my tongue their small skeletons made of rubbery limestone. . . . Suddenly, becoming as instinctive as fish . . . we turn . . . the veins of our blood which is marine. . . . Brushing the rocks with our faces. Smelling and kissing their mouths with our mouths.

> Facciamo parte del potente complesso marino. . . . Bevo un turbolento salvataggio di rizopodi. Ho sulla lingua i loro scheletrini di calcare gommoso. . . . Ad un tratto divenuti istintivi come pesci . . . ci voltiamo . . . le vene del nostro sangue che è marino. . . . Sfiorare le rocce colla faccia. Fiutarne e baciarne le bocche colle nostre bocche. (Marinetti 2003a, 46)[17]

In the second story, Ahmed Bey's futuristic desire for liberation from space and time is completed only after he commits suicide by jumping off the *Massilia* as it sails through the "carnal tenderness" (tenerezza carnale) of Guanabara Bay of Rio de Janeiro, and his corpse is maternally embraced by and merged with a comforting "radiant crimson Bosphorous with golden fringes of Brazilian ocean that wrapped him all over" (Marinetti 2003b, 109; sfolgorante Bosforo di porpora con frange dorate d'oceano brasiliano che lo fasciava tutto).

To be clear, if read from an ecofeminist standpoint, the feminization of nature and the subsequent nature-merging images in these passages are likely to be interpreted as further proof of an androcentric stance that perpetuates the possession and domination of women, animals, and the natural world rather than the manifestation of a sincere desire to nonhierarchically level and actualize the male narrators' self with those same entities. Especially if one thinks about Val Plumwood's (1993) conception of the "expanded self" as one that implicitly denies the relevance of other individuals (i.e., women) in order to affirm a nebulous, undifferentiated "cosmic" (male) self, the position Marinetti takes with his protagonists certainly remains more conservative than it initially appears.

Nonetheless, Marinetti does seem to occasionally use the technology and machines at his disposal (including the *techne* of his writing) to praise and valorize nature rather than merely conquering or taming it. For example, in the chapter "Il bosco di Provenza ove nacque il manifesto 'Uccidiamo il Chiaro di Luna'" (The Provencal forest where the manifesto "Let's Murder the Moonshine" was born), part of his autobiographical *Una sensibilità italiana nata in Egitto* (An Italian sensibility born in Egypt), the poet informs us that it is only when his fancy and powerful car, "Isotta Fraschini one hundred horsepower" (Isotta Fraschini cento cavalli), breaks down and he is forced to stop "in a melodious and resin-flavored Provençal forest whose shadows watermarked by blue green rays roused many nightingales to sing" (in un melodioso e resinato bosco di Provenza le cui ombre filigranate di raggi blu verdi eccitavano al canto molti usignoli) that he is able to "fix in his notebook the most original images of the famous manifesto" (1969b, 227; fissare nel [suo] taccuino le più originali immagini del famoso manifesto). A passage in the chapter "Verdi orli palpitanti di Milano" (Green throbbing edges of Milan) in his other autobiographical work *La grande Milano tradizionale e futurista* (The great traditional and futurist Milan) conveys a similar idea. When the poet finds himself outside the city in the green belt around Milan formed by the "countryside around Bergamo and Como" (1969a, 155; le campagne bergamasche brianzole e comasche), he tells Jesus that he is in this place to make a humble offering:

> We wandering artists are here offering the little we have that is art verses paintings dreams prayers of streams azure sculptures of incense.... Before opaque beings come to trample on us we must well extol the fluent contacts of the best Italian landscapes.
>
> Siamo qui artisti raminghi offrendo il poco che abbiamo cioè arte versi pitture sogni preghiere di ruscelli sculture d'incenso azzurrine.... Prima che esseri opachi vengano a calpestarci dobbiamo bene esaltare i fluidici contatti dei migliori paesaggi italiani. (155–56)

One cannot but wonder about this apparently polemical reference to such "opaque beings," and it is tempting, even though unlikely, to imagine that Marinetti was foreshadowing the devastating increase of illegal construction in Italy during the economic boom in the fifties.

More often than not, Marinetti's and futurism's attraction to a specific landscape or geographic area translates into a desire to incorporate it and literally eat it. While it is true that a voracious, possessive attitude toward reality—one that seeks to merge the self with the surrounding world and betrays a fundamental wish for self-fulfillment—characterizes most of his literary representations of technology-mediated relationships with landscapes, a distinction could be made based on whether such landscapes are in Italy or in exotic colonial locations such as Africa or South America. If, in the latter instance, "the combination of incorporation and abjection . . . exposes the violent, split unconscious of orientalist discourse" (Palumbo 2003, 7), the references to local and indigenous landscapes become an occasion to patriotically celebrate and accentuate the unappreciated natural beauties of the mother country.[18] In this sense, although authored by Fillìa (alias of Luigi Colombo) and not Marinetti, two recipes among "Crucial Futurist Meals" in their jointly written *The Futurist Cookbook* (1989) illustrate this sort of oscillation between a rhetoric of incorporation and one of celebration.

In a different context, "Geographic Meal," also in *The Futurist Cookbook*, replicates the familiar imperialist, consumptive, and male chauvinist pattern Blum (1996) discusses in relation to Marinetti's African texts. It describes a site-specific meal ordered by touching the different body parts of a young, shapely waitress tightly wrapped in a map of Africa. The woman thus functions as a living menu and edible matter, with the continent literally molding onto and becoming one with her body:

> If a table companion points a finger on the left breast of the waitress-menu where CAIRO is written, one of the waiters . . . will return . . . bringing the food corresponding to that city. (Marinetti and Colombo 1989, 186)

> Se un commensale punta un dito sulla mammella sinistra della cameriera-listavivande dove è scritto CAIRO, uno dei camerieri . . . ritornerà . . . portando la vivanda corrispondente a quella città.

In another chapter, "Meal Synthesis of Italy," the trope of incorporation functions as a means for Italians to proudly reclaim and celebrate—through all their senses—the geographic and alimentary variety of their mother country:

> In the past, Italy has always been a delicious food for foreigners. Today we can taste it but, if we wish to try at the table the flavor and smell of all its orchards, its pastures

and its gardens, we [still] are not able to be served all at once its many regional foods. (Marinetti and Colombo 1989, 183)

> L'Italia è sempre stata, nel passato, un ghiotto cibo per gli stranieri. Oggi possiamo gustarla noi, ma non ci è dato, volendo provare a tavola il sapore e il profumo di tutti i suoi orti, i suoi pascoli e i suoi giardini, di farci servire in una volta sola le tante vivande regionali.

More specifically, the goal of this lunch is to merge the outside, as is reproduced by different landscape pictures made by famous futurist painters, with the inside. As a consequence, this merging would produce an overwhelming multisensorial bodily experience that allows a simultaneous enjoyment of the indigenous foods and the territories that produce them:

> an alpine landscape Depero . . . a volcanic landscape Balla—a southern seascape animated with islets Prampolini. Before eating, the guests paint their hands with Mitilene blu. (183)

> un paesaggio alpino Depero . . . un paesaggio vulcanico Balla—un paesaggio di mare meridionale animato d'isolotti Prampolini. I convitati, prima di mangiare, si tingono le mani con blu di Mitilene.

Futurism's perception of landscape as a dimension that mediates between outside and inside, self and world, and subjectivity and objectivity finds its highest and yet most ambiguous achievement with "aeropoetry," Marinetti's creative intention to celebrate the sensorial experience of flight through poetry (and through painting with "aeropainting"). As Daniel Mangano observes, there is a definitive ambiguity at the base of aeropoetry because if aeropoetry is a way to penetrate the atmosphere's "hard thighs" (Marinetti 1990f, 1097) and thus the substance of the air through the poet's fusion with the machine, it is also an upward shift and thus a new distancing of the world (Mangano 1996, 60–61). Particularly relevant to my discussion are Mangano's remarks that Marinetti and other futurist aero poets—such as Gaetano Pattarozzi, Bruno Sanzin, and Paolo Buzzi—were able to metaphorically amalgamate landscape, machine, and poetic self to reduce the distance between the world and the "I," bringing the latter closer to the "cosmic dimension of matter" (61).[19] Futurism's intuition that subjectivity and objectivity merge through the mediation of landscape would seem to indicate at least an aesthetic interest in the environment, whereby human beings and the world in which they live would symbiotically aim to coexist in a new, in-between artistic reality that is the product of neither art nor the environment alone. But there is more. The idea that subjects and objects, bodies and things, organic and inorganic matter belong to—and have an impact on—the same natural, universal dimension strongly resonates with some postmodern ecocritical theories that investigate the extension of human agency, the boundaries of human and nonhuman, nature and culture, and in short, the intertwinements "of bodies, natures, and meanings" (Iovino and Oppermann 2012, 450).

Futurism's fascination and "lyrical obsession" (Marinetti 1990j, 50; ossessione lirica) with the concept of matter is well known. Christine Poggi observes, for instance, that matter "allows Marinetti to confound the organic and the mechanical, the procreative, and the industrial, and thereby to seize for himself the illusory power of male autogenesis" (1997, 23). The final paragraph of the dedication in Marinetti's *Mafarka the Futurist* well illustrates her point. In this passage, which anticipates the fantastic birth of Gazurmah—a superhuman, handmade being industrially produced without a woman—Marinetti announces that "men . . . with steel chins will prodigiously give birth, just by extending their exorbitant will-power" (1998a, 3; uomini . . . dal mento d'acciaio figlieranno prodigiosamente, solo con uno sforzo della loro volontà esorbitante) and that "men's spirit is an unused ovary" (3; lo spirito dell'uomo è una ovaia inesercitata) that will be fertilized by the futurists. Foreshadowing the figure of a cyborg, Gazurmah's formidable, airplane-like body is made with heterogeneous materials that combine the human, the animal, and the vegetal—(i.e., "palm fibers," "cloth," "flexible splints," "cat guts" [160; "fibre del palmizio," "tela," "stecche flessibili," "budella di gatto"])—as if such a combination were essential to the advent of a "multiplied" human being. As Jeffrey Schnapp reminds us, "The overcoming of physical decay by means of the forging of new bodies and materials had always figured among Futurism's heroic themes. . . . New materials like high-speed steel, aluminum, zinc-aluminum alloys, tempered glass and plastics became both emblems of a crystalline modernity that had emerged from out of decadence's dark shadows, and the body double or prosthetic extension of the new multiplied man and woman" (2004, 230–33). This plural and more-than-human entity, in short, emerges out of a peculiar techno-primitivist production process that combines organic and inorganic materials.

In what follows, I thus consider this futurist obsession with matter in conjunction with the recent development within ecocritical theory known as the material turn. As discussed earlier, the latter has expanded ecocriticism's practical applications "beyond nature writing" and "beyond nature" to a vision associating nature with human-centered concepts such as the "other-than-culture," "wilderness," or "the environment" (Iovino and Oppermann 2012, 449). Since this materialist ecocritical approach does not exclusively put nature, landscape, or the human at the center of the discourse, it is a more appropriate lens through which to examine an apparently hostile context such as futurism from an ecocritical perspective. In particular, I am interested in assessing to what extent Marinetti's representation and understanding of the multifaceted concept of matter may be contributing to the discussion originated by one of the fundamental questions posed by postmodernist material ecocriticism: "How material are we?" or, better yet, "How material-discursive are we?" (454).[20]

It goes without saying that the body—as a privileged, environmentally interactive playground for matter's many narratives—remains central to such a question. Yet my main objective is not to provocatively extract Marinetti from his historical context and transform him into a precursor of posthuman ecocritical thought.[21] Rather, I seek to draw attention to certain affinities, parallelisms, and "adjacencies" (Deleuze and Guattari 1986, 7–8) between the way futurism imagined matter and some of the current positions of material ecocriticism.

Of course, the fundamental reasons behind futurism's and posthuman ecocriticism's common interest in matter are substantially different. In simplified terms, in the former case, such an interest is an ideal means to defeat time and the perishability of the organic body through the abolition of the "I" and the advent of a strengthened "multiplied man." In the latter case, it is a way to question precisely the primacy of bio-, logo-, and anthropocentrism and to become aware of our nonhierarchical interrelatedness with a nonhuman universe. Additionally, I am aware that (1) in Marinetti, the term *matter* derives from Bergson's philosophy, it refers to the totality of a subject's perceived images, and it does not exclusively coincide with the physical reality of objects, and that (2) as Paolo Valesio (2013) observed, Marinetti was an intense poet of matter, but he never was a materialist in the radical and philosophic sense of the term. As Roberto Terrosi notes, "The main tendency of futurism is not to negate man or the human but to magnify it . . . thus maintaining a thread of continuity with the positions of humanistic renaissance" (2009, 268). At the same time, however, Bergson's "Darwinian view of mind as 'thinking matter'" represents also a "challenge to basic tenets of Enlightenment humanism" (Raine 2014, 109)—a challenge that futurism certainly made its own.

This said, despite their clear ideological divergences, futurism's and ecocriticism's respective explorations of the interrelations of human corporeality with a universe constituted by other-than-human elements illuminate each other. Specifically, the two discourses share a concern with the alchemical force and vibrant life of things and a desire to register the agency and the stories of both human and nonhuman entities. Furthermore, both positions have a common interest in the precarious and not-always-central position of human beings in a rapidly transforming world as their underlying philosophies strive to expose a continuity between the organic and the inorganic of bodies and things, of beings and objects. I am intrigued when, for example, in her *Vibrant Matter*, Jane Bennett states that her aim is to stress the active role of nonhuman materials and thing-power in public life, thus inviting her readers to develop the naivete and sensibility to the outside that are associated with vital materialism. My attention is similarly stirred when, within the same discourse, Bennett encourages readers to rediscover the theoretical wealth of premodern philosophies of nature such as animism, vitalism, and anthropomorphism. And finally, some of futurism's own tales and positions keep "dangerously" coming to mind when Bennett asks us to direct our attention to tales that, both ambitiously and naively, ask questions "that have been too readily foreclosed by . . . the fetishization of the subject, the image, the word" (2010, 19).

In particular, I cannot help but think of Marinetti's many "naive" remarks regarding the need to put an end to the superiority of the human, to "penetrate the essence of matter and destroy the dull hostility that separates it from us" (1990j, 52; penetrare l'essenza della materia e distruggere la sorda ostilità che la separa da noi), and to operate on the discursive level of the text so that "literature enters directly the universe and becomes one with it" (53; la letteratura entri direttamente nell'universo e faccia corpo con esso). I think too of Giacomo Balla and Fortunato Depero's playful intention in their "Ricostruzione futurista dell'universo" (futurist reconstruction of the universe) to provide "skeleton and flesh to the invisible, the insubstantial, the imponderable, the imperceptible" (1915; scheletro e carne all'invisibile, all'impalpabile, all'imponderabile, all'impercettibile), thus articulating what Bennett identifies as their desire to infuse a corporeality and a vital materiality into the quasi body of a universal vibration (2010, xiii).[22]

The language of these few futurist excerpts shows once again how crucial the body is in this context and how the expansion (or reduction) of its perception may resemble some of the positions of material ecocriticism. In other words, the notion of corporeality loses its familiar, fixed association with the human specifically. Instead, it becomes a dynamic process that transits, is diffused, and defines other potentially sentient entities beyond the human. In this scenario, the nonhuman can also be embodied. Such nonhuman embodiments may include a text that must now be integrated with the outside world (i.e., "to make a body") or even an invisible vibration in the matter of the universe that acquires a skeleton and flesh.[23]

There are many invitations in the works of Marinetti and other futurists to embrace "the same life of matter . . . in order to provide the maximum amount of vibrations and a more profound synthesis of life" (Marinetti 1990a, 73–76; vita stessa della materia . . . allo scopo di dare la massima quantità di vibrazioni e una più profonda sintesi della vita), to diminish the ontological distance that separates human beings and things, and to represent what Bennett calls the "vitality intrinsic to materiality" (2010, xiii).[24] Remarks such as "The sand lives and thinks, but doesn't want to speak" (Marinetti 1990c, 1075; Vive e pensa la sabbia, ma non vuole parlare) in *Il fascino dell'Egitto* or "Fabrics fabrics fabrics fabrics fabrics fabrics and since I am you all are sensitive thinking speak at last" (1990d; Stoffe stoffe stoffe, stoffe stoffe stoffe e poiché sono siete tutte sensibili pensanti parlino parlate parla finalmente) in "Il poema non umano dei tecnicismi" could be viewed as attempts to establish a continuity between the human and the more-than-human dimension, as they assign a thriving material agency to substances in the physical world, be it the sand of the Egyptian desert or a new industrially produced textile.

Notably, building on intuitions that parallel mine and generally drawing from a similar theoretical background, Danila Cannamela devotes her remarkable *The Quiet Avant-Garde* to a detailed exploration of the destabilizing confusion between people and things that emerges both in the poetics of the futurists and, even earlier, in one of the proto-avant-gardist crepuscular poets. Paying attention to the wide-ranging implications of such a blurring, nonanthropocentric gesture, her discussion addresses a large number of revealing texts from these literary movements that invite us to rethink "our sense of humanity in relation to the nonhuman environment" (2019, 10). As such, her book represents a useful complement to my observations in this section.

My initial focus here is on Marinetti's (1960a, 1960c) plays, which aim to represent a series of animated, sentient objects removed from their traditional setting and situate them in an unusual condition. Once creatively repositioned, their astonishing essence, construction, and nonhuman life emerges, and they stand out as protagonists in some of the clearest texts in which, ecocritically speaking, action is generated by the mutual collaboration between the human and the nonhuman.

In the short play *Vengono* (They are coming; 1960c), for example, the lives of the human and nonhuman characters—the majordomo and the servants, and the pieces of furniture (an armchair, a table, and eight chairs), respectively—are ideally interconnected, since they belong to the same existential horizon and have equal performative standing on the stage's dynamic topology. Their interactive engagement as they move and are moved around, waiting and listening together for some mysterious guests to arrive, is the real source of dramatic tension and shapes the piece itself. As Marinetti writes, "I wished to create a synthesis of animated objects . . . the spectator . . . must feel that the chairs really live and move by themselves to get out" (1960c, 285; Ho voluto creare una sintesi d'oggetti animati . . . lo spettatore . . . deve sentire che le sedie vivono veramente e si muovono da sole per uscire). A similar intention to "convey the nonhuman life of objects" (1960a, 345; dare la vita non-umana degli oggetti) is found in *Il teatrino dell'amore* (The little theater of love). Here the pieces of furniture, a cupboard and a sideboard, even possess voices, and they exchange onomatopoeia-rich lines:

> CUPBOARD: Cric. It shall rain in three quarters of an hour.
> SIDEBOARD: Crac-crac. On the third floor the servant is going to bed.
>
> IL BUFFET: Cric. Pioverà fra tre quarti d'ora.
> LA CREDENZA: Crac-crac. Al terzo piano la serva va a letto. (342)

Together with a little girl, they witness from their apparently wiser existential condition of thingness a depressing bourgeois scene of human love and betrayal between the girl's mother and "the first guy to come along" (342; il primo venuto).

Marinetti's explanation of what happens in this play is particularly interesting, since he specifically points out the nonhuman origin of the vitalistic impulse that characterizes the objects on the stage:

I wished to convey the nonhuman life of objects. The most important characters are the little wood theater (whose puppets act in the dark without the presence of the puppetmaster), the cupboards, the sideboard, and they are not humanized (like sometimes other things were in the past theater) but nonhumanly convey the temperature, their expansions, the weight they can bear, the vibrations of the walls, etc.

Ho voluto dare la vita non-umana degli oggetti. I personaggi più importanti sono il teatrino di legno (le cui marionette recitano nel buio senza la presenza del burattinaio) il buffet, la credenza che non sono umanizzati (come qualche volta furono umanizzate le cose nel teatro passatista) ma danno non-umanamente la temperatura, le loro dilatazioni, i pesi che sopportano, le vibrazioni dei muri, ecc. (1960a, 345)

In other words, the vitality and animation of the inorganic are not just the result of a process of humanization or simple anthropomorphism but are inherent to matter itself, to the materials (i.e., the wood, metal, and paint) that constitute the bodies of these active objects, which are also becoming speaking subjects and "actants" in the play.

Thus in these short experimental pieces, we see that "the old conceptions of matter as stable, inert and passive physical substance, and of the human agent as a separate observer always in control, are being replaced here by . . . models that effectively theorize matter's inherent vitality" (Iovino and Oppermann 2012, 465). In these examples, matter is not just "a vector of speed, a volatile, masculine substitute for . . . nature" (Poggi 2009, 156) and another instance of Marinetti's familiar animistic tendencies. It is also not merely something that helps the futurist's fight to defeat time and combat "against rot" (Schnapp 2004, 230). Rather, Marinetti's interest in matter reveals both a pseudoscientific, esoteric fascination with "the incomprehensible and inhuman alliance of its molecules or its electrons" (Marinetti 1990j, 50–51; l'alleanza incomprensibile e inumana delle sue molecole o dei suoi elettroni) and a bent for a poetic, even mystical attitude of reenchantment with, and wonder about, the nonhuman world. These different perspectives ultimately depend on us: we might choose to emphasize and view futurism through either a "theosophic" (Cigliana 2002, 237), premodern lens or an "onto-poetic" (Iovino 2012c, 142), postmodern, material-ecocritical one.

Although in the previous texts, matter possesses what Bennett calls the "quivering effervescence" of life (2010, 55), there is no doubt that it is precisely matter's fundamental inorganic and other-than-human quality that fascinates futurism. The "fabrics of modern times" discussed by Schnapp in Marinetti's

"Il poema non umano dei tecnicismi," together with a number of new modern, industrially produced materials from steel to plastics, helped energize and awaken such an individual body from its decadent torpor and, as mentioned earlier, functioned as a way to enhance and multiply the future human race.

Leaving aside fabrics and adding some iron and concrete, a text like Marinetti's *Ricostruire l'Italia con architettura futurista Sant'Elia* (Reconstructing Italy with Sant'Elia's futurist architecture) suggests that those same materials are considered essential to reinvigorating the nation's body politic and health, especially when some of its parts, like the city of Venice, are thought to be weakened and dying because of their association with the past and thus in need of being destroyed and then rebuilt according to futurist principles. Marinetti's architectural "divertimento rappresentabile in molte sintesi" (1960b, 507; divertissement representable in many synthesis) thus promises to generate some additional reflections on the futurist interactions among nature, bodies, and matter, as it expresses concern for the future of the Italian territory in general and the city of Venice in particular.[25]

Inspired by the polemical debate between rationalists and futurists at the end of the twenties on the objectives and qualities of modern architecture, *Ricostruire l'Italia* dramatizes the different and yet ultimately overlapping positions of the "Speedists" (Velocisti), who stand for the futurists, and the "Spatialists" (Spaziali), who stand for the rationalists. These two conflicting groups eventually search for common ground, reconciliation, and collaboration against the threat represented by the "Softies" (Mollenti)—that is, the traditionalists—who are attached to old-fashioned conceptions of urban architecture and landscape planning. On the one hand, this work cannot but reconfirm its author's distance from any modern idea of environmental or ecological consciousness. Vasto, the leader of the Spatialists, makes this clear in the first synthesis, "Lo spazio vivente" (The living space), when he establishes a direct correspondence between the human and the national body:

> We, magnified men, need a magnified homeland. The peninsula will be augmented, at every tip of its edge, with 200 kilometers of reinforced concrete and iron that will cover the sea filled with the pieces of the razed-to-the-ground mountains.
>
> Noi, uomini ingigantiti, esigiamo una patria ingigantita. La penisola sarà aumentata, in ogni punta del suo orlo, di 200 chilometri in cemento armato e ferro che copriranno il mare riempito coi pezzi delle montagne rase al suolo. (Marinetti 1960b, 520).

Ballamar, one of the "spatial architects" (Marinetti 1960b, 511; architetti spaziali), echoes this position. While certainly not a supporter of the natural world, complaining as he does about the domination of "the line of the very beautiful sea" (520; linea del bellissimo mare) and outlining his plans to reshape it "through new kinds of waves" (523; mediante nuovi tipi di onde), he is also the proponent of a fragile, ephemeral, primitive, theriomorphic architecture that, in a familiar dynamic, betrays a persistent fascination with nature's shapes and functioning principles:

> The frameworks are wings! . . . The buildings will have immense and sharp ostrich breasts and pirouetting balconies. The dams of the new harbors will have the synthetic shapes of waves and fish. . . . Expansive architectures made of diverse matter will unite with the clouds, the rain, the snow, the fog, the darkness. . . . Projectors, by weaving their beams of light into the clouds laid on the brow of the building, will complete the building itself with a splendid germination of luminous crystals.
>
> Le armature sono ali! . . . Gli edifici avranno smisurati e taglienti petti di struzzo e terrazze piroettanti. Le dighe dei nuovi porti avranno le forme sintetiche delle onde e dei pesci. . . . Architetture espansive e polimateriche si uniranno alle nuvole, alla pioggia, alla neve, alle nebbia, alle tenebre. . . . I proiettori, intrecciando i loro fasci di luce nelle nuvole coricate sulla fronte dell'edificio, completeranno l'edificio stesso con una splendida germinazione di cristalli luminosi. (523–24)

Both Ballamar and Vasto, as well as the leaders of the Speedists (Furr and Vif-Glin), are confident in the powers of new construction materials. Their brief architectural debates revolve around the supposed eternity or ephemerality such materials should possess, and they eventually compromise on the mutually agreeable, paradoxical solution of an "ephemeral spatial concrete" (Marinetti 1960b, 524; cemento spaziale effimero). As Vasto says,

> I settle for the eternal dressed with a variable ephemeral. . . . In order to imitate life we shall build on granitic bases all the disproportioned, unequal elasticity of ephemeral concrete. . . . It is solid. We shall destroy it when it will bore us. Above and below it, marble, basalt and stones will resist.
>
> Concludo per l'eterno vestito di effimero variabile. . . . Per imitare la vita costruiremo su basi granitiche tutte le elasticità sproporzionate inegualiste del cemento effimero. . . .

È solido. Lo distruggeremo quando ci annoierà. Sopra e sotto di lui il marmo, il basalto e le pietre resisteranno. (528)

What is particularly interesting in this passage is not only the deliberate effort to merge the inorganic aspect of the constructions with the organic one of the animals—to unite artificial and natural matter—and the fact that architecture becomes here a metaphor for human existence but also the inherent vitality and corporeality of buildings that possess wings, breasts, and a forehead. The limited time span of concrete, its eminently lifelike quality, and especially the implication that the construction materials possess a dormant vibrancy of their own should also be highlighted. This vibrancy, once energized with the combination of natural elements (clouds, rain, etc.) and artificial ones (electric light) will contribute to an almost auto-genetic completion of the buildings themselves.

Allusions to the possibility of an embodied life (and death) of inorganic matter and, vice versa, of a materialization of organic life also emerge in other moments in the play: Altaluce (Vasto's daughter), for example, remarks that "iron and concrete have been long intoxicated with the thought of killing themselves" (Marinetti 1960b, 533; il ferro e il cemento sono da tempo ebbri di suicidarsi); Alata (Vasto's wife) refers to Venice as a "corpse of a city" (542; cadavere di città); the title of the "Fifth Synthesis" is "La morte dell'ultimo treno" (The death of the last train); and finally, Vasto, Alata, and Furr are temporarily transformed into "living statues" (561; vive statue), emblematically chained by an oxymoronic assemblage of inorganic and organic matter constituted by some "very hardened vegetable chains" (571; induritissime catene vegetali).[26]

Beyond the surreal plot and self-serving propagandistic turn of the events depicted, what also animates *Ricostruire l'Italia* is the idea that human beings, habitats, and materials (i.e., bodies, nature, matter) share an interrelated inner vitality. As a consequence, this text foregrounds the interactions between different kinds of living bodies and their entanglement with the more-than-human world in Marinetti's work.

Moving away from the Northern, urban scene of Venice, in the final part of this chapter, I reflect a bit more on futurism's concept of nature and its engagement with matter in the context of a rural Sicilian landscape at the foothills of Mount Etna. Incidentally, the backdrop of Marinetti's play *Vulcano: 8 sintesi incatenate* (Volcano: 8 chained syntheses; 1960d) is not too far from the location of Pirandello's "Il fumo."

In spite of futurism's notorious fascination with technology and progress, the raw power and primitive, incendiary energy of volcanoes fascinated Marinetti since the beginning of his artistic career.[27] As Mirko Bevilacqua reminds us, "The volcano is the center of futurist theatre, the large circle of fire and [the] eruption of Marinetti's futurist evenings" (1977, 260). Unsurprisingly, in Marinetti's *Le monoplan du pape* (The pope's monoplane; 1912), two chapters ("Les conseils du volcan" [Suggestions from the volcano] and "Chez mon père, le volcan" [At my father's, the volcano]) are dedicated to Mount Etna. Here Marinetti establishes a significant parallel between the violent, natural pyrotechnics of the mountain and the disruptive, vertical, fictional powers embodied by futurist poetics. Mount Etna especially attracts him for its "unending plastic activity [and] for its eschatological vitality [situated] between the human and the divine" (Bevilacqua 1977, 260). Bevilacqua quotes a specific verse from *Le monoplan* to corroborate his observation: "I am incessantly (con)fused with my scorias / My life is the perpetual fusion of my debris" (260; Je suis incessamment mêlé a mes scories / Ma vie est la fusion perpétuelle de mes debris). This verse, he observes, represents the "superman's starting point" (260), as it alludes to the superhuman goal the protagonist aims at reaching with the help of both technology and nature (i.e., the "monoplan" and the "volcano," respectively). However, as Marinetti conveys an image of the self as a son of the mountain—a cocreating, hybrid, volcano-like accumulation and amalgam of human, mineral, and divine debris and scoria—he also hints at a quintessential transcorporeal and posthuman scenario.

Fifteen years after the composition of *Le monoplan*, Marinetti's *Vulcano* continues to play with the idea of a material continuity between the organic and the inorganic, the human and the nonhuman. At the same time, it also conveys a more problematic, skeptical look at the combined power of human beings and technology as efficient means to enact change or initiate any kind of rebirth, be it aesthetic, spiritual, or existential. Tangentially, Marinetti's attitude in this text resembles his ambivalence in *Le Roi Bombance* (1905), another early play in which he conveys a self-critical and pessimistic assessment of some of futurism's traditional strongholds—namely, the force of creative willpower and technology to transform reality and to reach for the "absolute."

In *Vulcano*, for instance, nature is represented in ambiguous ways. Depending on the characters' perspectives, it either possesses a potential regenerative power or does not seem to have any power or transcendental value at all. For example, according to Giovanni, a scientist with a fundamentally rationalist, empirical perspective on reality, "a volcano is a volcano, that is to say nothing literary, sentimental, or political" (Marinetti 1960d, 174; un Vulcano è un vulcano, cioè nulla di letterario né di sentimentale né di politico). However, for the poet, or for the sensually and spiritually inclined husband and wife aspiring to elevate their love, the volcano's lava is a sacred substance sent by God to "warm, embellish, rejuvenate, purify our Earth" (202; riscaldare abbellire ringiovanire purificare la nostra terra).

To complete the list of characters and briefly outline the plot, let us add that the main protagonists are two couples who, as hinted above, have very different positions and attitudes toward reality and nature. The first couple, Mario and Eugenia Brancaccio, are wealthy Sicilian landowners and entrepreneurs interested in "volcanic matters" (Marinetti 1960d, 143; materie vulcaniche). Their life of luxurious wandering is devoted to finding ways of merging with, if not literally disappearing into, the landscapes they encounter as they travel and, in the process, attempt to rekindle their love. In the following remarks, Mario poetically mixes the language of science and desire to suggest that human bodies and nonhuman substances are subject to the same universal rules and expressive urges and that there is a sensuous relationship between people and their surroundings. As he evokes the intertwining of interiority with exteriority and the merging of feelings, places, and matters, his words might well anticipate Maurice Merleau-Ponty's ecological vision of continuity between bodies and the world, "foreground[ing] the interconnectedness of human and nonhuman beings and phenomena" (Raine 2014, 107):

> Undoubtedly, there are laws that govern the emotions of souls and bodies, their fiery merging and the continuity of their feelings, [these] laws are similar to volcanic laws and perhaps they are influenced by them. Nothing can be beyond this kiss but death and the liquefaction of our bodies in the lava.... We would forget the last village we enjoyed and attempt to get acclimatized in a new one, intermingling our intimacy with the most typical part of the environment we explored.... In medieval cities... we merged our soul with their atmosphere of sick, moaning and tired bronze.

> Vi sono indubbiamente leggi che governano le simpatie delle anime e dei corpi, le loro fusioni ardenti e la continuità dei loro sentimenti, leggi simili alle leggi dei vulcani e forse influenzate da loro.... Al di là di questo bacio non vi può essere che la morte e la liquefazione dei nostri corpi nella lava.... Dimenticavamo l'ultimo paese goduto e tentatavamo di acclimatarci nel nuovo mescolando subito la nostra intimità con

la parte piú tipica dell'ambiente che noi esploravamo. . . . Nelle città medievali . . . abbiamo fusa la nostra anima con la loro atmosfera di bronzo malato consunto e gemente. (Marinetti 1960d, 175–77)

Mario and Eugenia's main objective is to get entangled with the world that surrounds them, and the overall scenario depicted is one of constant becoming and transformation—one "in which bodies [and souls] extend into places and places deeply affect bodies" (Alaimo 2016, 5). While I would be reluctant to affirm that Mario willingly and self-consciously "advocates pleasurable modes of environmentally oriented habitation" (13), his fundamental erotic register suggests that the notion of pleasure—a biophilic kind of pleasure that seeks to achieve some form of satisfaction by simultaneously engaging with human and nonhuman matter (namely, with Eugenia and the landscape at once)—is likely at play in this context. In short, being intimate here and truly alive, even after death, means to engage and dissolve (literally, "liquefy") together with many different bodies, be they human, mineral, lithic, or metallic.

The other couple, volcanologist Giovanni Massadra and his wife, Lucia, have a different main concern: to save people from "nature's tremendous whims" (Marinetti 1960d, 174; capricci tremendi della natura). They plan to do so by building a "lava-stopping machine" (202; macchina fermalava)—a mechanical device supposedly able to stop the erupting lava from destroying everything in its path by freezing it. Two additional players are Alberto Serena and Porpora. The former is a drug-addicted poet and war hero who finds himself in a creative crisis, has lost "faith in [his] own body" (156; fede nel [suo] corpo), and is no longer able to find the right words for his poetry: "I need a new red, hot word. . . . I cannot find it" (167; Ho bisogno di una nuova parola rossa calda. . . . Non so trovarla). The latter is a senile pyrotechnician (pirotecnico) who, despite his efforts to launch the most spectacular and colorful fireworks show, ultimately emerges as incompetent and unable to artificially replicate—much less surpass—the effects of a real volcanic eruption ("Damn! Damn! My face got burned!" [150; Maledizione! Maledizione! Ho la faccia bruciata!]).

As these brief descriptions suggest, *Vulcano* does not display the bellicose polemical rhetoric and revolutionary intents that are present in *Le monoplan*, in which a fatherly, anthropomorphic Mount Etna encourages through a sort of fire baptism Marinetti's new aesthetic and political program. On the contrary, the Sicilian mountain in *Vulcano* provides the natural backdrop for a series of successive failures affecting all professions and levels of society. From rich capitalists like Mario and Eugenia, to scientists like Giovanni and Lucia, to artists in crisis like the poet and Porpora—all are prevented from any sort of regenerative or progressive dynamic, be it aesthetic, scientific, or other. Regardless of their beliefs and stances vis-à-vis nature, poetry, and technology, all the various competitions and challenges (artistic or scientific) the various characters engage with as they attempt to artificially exploit, reproduce, surpass, or contain the volcano's power are eventually lost. The "elemental" death of Mario and Eugenia by volcanic lava and fire is the expected conclusion of this process. However, this event also represents the culmination of their erotic embrace and, as such, could be interpreted as a metaphorical reignition and rejuvenation of their passion, since they are now literally fused with each other and the world, to which they had aspired all along. In this sense, it resembles the sort of immanent "vital death" theorized by Rosi Braidotti in *The Posthuman*—namely, a human yearning "to disappear by merging into [a] generative flow of becoming, the precondition for which is the loss, disappearance and disruption of the atomized, individual self" (2013, 134). At the same time, it also reaffirms Mario and Eugenia's fundamental human insignificance and powerlessness against the volcano.

At the end of the play, the mechanical device invented by scientist Giovanni Massara temporarily stops the flow of lava but soon breaks down, showing all its limitations.[28] Then the carbonized corpse of Eugenia is found stuck and "fused with the metal itself" (Marinetti 1960d, 205; fuso col metallo stesso) inside his device, thus indicating both Marinetti's rethinking of the power of technology and his continuous reassessment of nature's ontological role. As he concludes *Vulcano* by depicting a situation in which molten lava, human flesh, and pieces of metal are literally enmeshed with one another, he not only overcomes a traditional, romantic vision of nature as a fully separate entity—a "transcendent source of aesthetic, spiritual, or moral value" (Raine 2014, 103)—but also hints at a fundamentally posthuman, transcorporeal scenario. As *Vulcano* complicates previous representations of the natural world, assigning to nature a powerful agency, and entangles human, mineral, and mechanical corporealities in a new hybrid, transcorporeal amalgam, it addresses "intriguing new ways of imagining the more-than-human world" (105). Furthermore, and perhaps more unexpectedly, it also resonates with Stacy Alaimo and Susan Hekman's observation that "nature 'punches back' at humans and the machines they construct to explore it in ways that we cannot predict" (2015, 146).

While I am obviously not suggesting that *Vulcano* is even remotely informed by any sort of environmental feminism, the play nonetheless invites us to think of less familiar forms of interaction between humans and the nonhuman. Ultimately, while this text questions a fixed and stable notion of the subject by exposing it to a natural process of disidentification, simultaneously foregrounding an ontological plurality based on a relational kind of subjectivity, the play also problematizes canonical humanist outlooks and epistemologies.

Once again, futurism's main interests and goals do not include realistic environmentalist concerns. Still, approaching a selection of Marinetti's works from an ecomaterialist perspective may invite us to (re)consider futurism's representation of nature, its thinking of the body, and its still unresolved relationship with the eloquence of matter. Because some aspects of the "vital materialism" discussed by Bennett (2010) can be traced to this modernist context, we are able to navigate these seemingly incompatible views on Marinetti.

Specifically, his work hints at the possibility of a nonhierarchical mutuality, interaction, and coexistence between human subjects and nonhuman objects, between bodies and things, and between two positions: on the one hand, interpretations that perceive Marinetti as someone who gives "always more importance to matter, marginalizing and diminishing human beings" (Curi 2009, 296), and on the other, those affirming the poetic and vital anthropocentric prevalence of the creating subject over matter and things (Ceccagnoli 2009, 329). To conclude, it is worth noting that when Marinetti wrote that it is necessary for literature to directly enter the universe and become one with it, he was referring to futurism's need to find a new kind of language and eliminate syntax. With a little imaginative effort, however, his aspiration might also be interpreted as aiming to establish a cultural (poetic) ecosystem in which the discursive and the material world are embodied together.[29] Here, if nothing else, the long-term survival (if you wish, aesthetic sustainability) of this ecosystem depends on humans sharing "their narrative horizon with . . . other things" (Iovino 2012a, 66). In other words, the notion of a corporeality that is no longer exclusively human or organic but is also material finds one of its earliest and most intriguing manifestations in Marinetti's modernist work.

2

Slick Territories
Petroculture, Italian Style

Gasoline is divine.

La benzina è divina.

—F. T. Marinetti, "La nuova religione-morale della velocità"

The true, only God in Sarrok is him, Petroleum.... In Arasolè one could become an ox, but in Sarrok one will certainly become a screwdriver.

Il vero, unico, Dio a Sarrok è lui, il Petrolio.... Uno poteva diventare bue ad Arasolè ma diventerà sicuramente un cacciavite, a Sarrok.

—Francesco Masala, *Il parroco di Arasolè*

At the beginning of his seminal essay "Petrofiction: The Oil Encounter and the Novel,"[1] which initiated the ecocritical subfields of petroculture studies and petrocriticism, Indian novelist and cultural critic Amitav Ghosh famously wonders why, despite oil's omnipresent influence and value as a global commodity, its encounter with world literatures "has produced scarcely a single work of note" and has generally "proved so imaginatively sterile" ([1992] 2005, 138–39). As he reflects on this issue, he observes that "to many Americans, oil smells bad ... it becomes a problem that can be written about only in the language of solutions" (139); that "the silence extends much further than the Arabic- or English-speaking worlds"; and that the crucial reason for "the muteness of the Oil Encounter [mostly between the United States and the Middle East]" needs to be found in "the craft of writing itself" (141). By this, he means that there are some fundamental disjunctions between the world of oil, which he qualifies as multilingual, displaced, heterogeneous, international, and the narrower one of the novel, which is typically monolingual, evoking a specific "sense of place" and "reveling in its unique power to evoke mood and atmosphere" (142). His partial conclusion, which also helps him introduce and discuss *The Trench*, part of Abdelrahman Munif's cycle of "oil fictions," is thus that "we do not yet possess the form that can give the Oil Encounter a literary expression" (142).

More recently, scholars such as Imre Szeman (2012) and Graeme Macdonald (2012) have moved beyond Ghosh's position, arguing that the term *petrofiction* should not just refer to fictions explicitly about oil but rather be expanded to indicate those broad "energy-dealing" narratives created during a whole period ("petromodernity") that may stretch as far back as the early industrial era. Szeman in particular requests that writers create "narratives that shake us from our faith in surplus . . . not by indulging in the pleasures of end times or fantasies of overcoming energy limits but by tracing the brutal consequences of a future of slow decline, of less energy for most and no energy for some" (2011, 325). A similar invitation is implicitly shared by Jennifer Wenzel. After pointing out oil's paradoxical "ubiquity and invisibility" in our lives and its simultaneous metaphorical and material quality, Wenzel suggests that the central critical questions today should focus on the extent to which "different kinds of texts— novels, short stories, poems, manifestoes, essays, cartoons, photographs, and documentary films—either work against or contribute to oil's invisibility" (2014, 3). Last but not least, Stephanie LeMenager's *Living Oil: Petroleum Culture in the American Century* emphasizes the crucial interdependence between cultural forms (novels, films, music, etc.) and fossil fuels. LeMenager's focus on the multifaceted bioregional implications of oil on ecosystems, people, and animals (i.e., in California and the Niger Delta) and her intention to "materializ[e] the ecologies of modernity" (2014, 184) are especially inspiring and methodologically useful for my own place-based approach.

It is thus with these sorts of questions and concerns in mind, and in the wake of this increasing critical and theoretical attention to the still prevalent energy source of our era, that this chapter wishes to selectively explore Italy's own discursive and material encounter with oil from the early decades of the twentieth century, when it began its rise as a global source of energy, until today, when the implementation of alternative and more sustainable energy sources continues to be challenged and hindered by a global cheap-oil economics. Of course, some may think that, given the particular place and context of my exploration, this could be an unrealistic objective—literally, a waste of energy. After all, if it is true, as earlier criticism has claimed, that in oil-rich places like the United States or the Middle East, there is an "almost complete absence of oil as a subject matter (direct or allegorical) in literature written when it [oil] is dominant" (Szeman 2011, 324), what might one expect to find in the cultural production and geological depths of what is a minimally oil-producing and mostly importing western European country—one that certainly does not stick out in the global imaginary for the wealth of its energy resources and hydrocarbon reserves?[2] Karen Pinkus aptly notes that "oil is not produced in the Italian subsoil, at least not in any notable quantities.... Until World War I hydrocarbons represented a negligible percentage of Italy's energy use" (2012, 152). Nonetheless, as she later ponders on how crucial a reflection on fossil fuels is "to our survival" (2016, 4), some figures belonging to the Italian "cultural humus" (Agamben, Marinetti, Pasolini) surface in her intriguing speculations about energy.

Strictly speaking, it is rare to find petroleum as a topic or a direct referent in the modern and contemporary Italian literary tradition. However, such scarcity may easily change if, following these more recent scholarly directions, we track its more oblique representations. In other words, if we view the issue in terms of the "cultural dimension of oil capitalism," we pay attention to the broader "array of cultural forms that have propelled and, at the same time, resisted and/or protested the oil economy," and we ultimately consider "oil as both an industry and a culture, a business and a set of aesthetic practices, a natural resource and a trope" (Barrett and Worden 2014, xxi–xxii). Italy's role thus becomes just as significant as that of the Americas, the Middle East, or any other country depending on oil. And if we consider both the "exuberant" and "catastrophic" kinds of discourses that "have informed literary representations of oil's meaning since the nineteenth century" (Buell, quoted in Barrett and Worden 2014, xxvii)—where "catastrophe" paradoxically includes both discourses about resource scarcity and antipodal dialogues centered on the environmental consequences of oil's abundance—the number of cultural works to examine becomes daunting.

This critical operation is further enriched when, together with the "cultural," one also looks at the material dimension of oil capitalism, at how oil variously affects and permeates environments (i.e., refineries and gas stations, invisible by-products, and health hazards), and at how its imbrication with processes of nation building implicitly reveals the extent of our interdependence on and interchanges with the nonhuman world. To put it in more scholarly terms, when taking inspiration from material ecocriticism, we see oil as a text in itself and as an eloquent, generative force that simultaneously broadens "the narrative potentialities of reality" and those of "traditional texts like books and films" (Past 2016, 378). In a recent essay that methodologically parallels mine, Heather Sullivan further clarifies how material ecocriticism, by considering oil as a text, may help assess how the latter fuels an entire cultural ecosystem. She observes, for example, that "in petrocultures, a petroleum-derived road system carries food [that is] itself fueled and protected by petroleum-based fertilizers, herbicides, and pesticides, and transports it in vehicles using petroleum or other fossil fuels contained in petro-based plastics" so that finally, "petro-foods and petro-paths produce petro-texts" (2017, 414).

In the following pages, I focus my attention on a selection of what Sullivan defines as "petro-texts." First, I consider some Italian modernist texts that—referring back to Ghosh and Wenzel—found both the way and the form, even though fragmentary, to give an early, exuberant, promotional voice to the "oil encounter" and that, given the historical moment in which they were written, reflected and helped "establish the conditions necessary for the rise of a dynamic oil economy" (Barrett and Worden 2014, xxvii). I then venture more concretely into the literary and material "slick territories" I mentioned in this chapter title, inspired by a mixture of textual and visual examples that reflect oil's fluidity and ability to cross multiple mediatic forms. As the modernist infatuation with petroleum that characterized a good part of the twentieth century shifts toward more critical and challenging attitudes toward oil, the full range of petroleum's transcorporeal agency on bodies and environments will also become apparent.

Futurist Fuels

As Joshua Schuster states in his *The Ecology of Modernism: American Environments and Avant-Garde Poetics*, "Oil is everywhere during the modernist era, changing the shape of the landscape with cars, roads, airplanes, military equipment, spawning suburbs, intensifying land speculation and commodity trading, further mechanizing agriculture, and producing new chemicals and plastics" (2015, 162). With minimal adjustments, these words could easily refer to the European and, specifically, Italian situation during the first few decades of the twentieth century, when the futurist avant-garde was most active.

Futurism's well-known aesthetic and material fascination with machines and the objects of technological modernity (cars, planes, ships, tanks, electricity, speed) immediately signals how it is metaphorically drenched in oil and, unsurprisingly, enthusiastic about oil's potential to fuel and provide inspiration to the movement. This substance complements and epitomizes the futurists' "lyrical obsession" (Marinetti 1990j, 50) with matter and reflects their interest in investigating various kinds of entanglements between human bodies and nonhuman realities.

Notoriously, in "Fondazione e manifesto del futurismo" (Foundation and manifesto of futurism), the foundational text of the movement, Marinetti-the-driver needs first to emerge together with his car out of a muddy, oily ditch, with his "face covered with the good grime of mechanic shops—a mix of metallic debris, useless sweats, celestial soot" (1990b, 9–10; volto coperto della buona melma delle officine—impasto di scorie metalliche, di sudori inutili, di fuliggini celesti), before being able to declare and eventually publish the eleven points at the core of his nascent movement, as if such a greasy combination of human sweat and by-products of gasoline were crucial not only to drive a vehicle but also to lubricate his imagination and thus literally thrust futurism forward. The quintessential ecocritical idea that the world where artists conceive their works (be they literary or otherwise) is the same one in which hydrocarbons, in one way or another, affect everyone's lives and creative activities in a "global system in which energy, matter and ideas interact" (Glotfelty 1996, xix) could hardly find a clearer or more emblematic expression.

A similarly revealing reference to oil surfaces in Marinetti's "La guerra elettrica" (The electrical war), one of the sections of "La guerra, sola igiene del mondo" (War, the world's only hygiene; 1990g). This text begins as a utopia by depicting a "futurist wonderland, where famine and the drudgery of labor have been conquered through mechanized agriculture" (Blum 1996, 43) but approaches its end with the hellish, dystopian—not to mention, from a contemporary perspective, eerily prophetic—picture of a global electric war among twenty-five superpowers to control their surplus production. This futurist "vision-hypothesis" (visione-ipotesi) is anchored in the central image of a feminized earth that is coextensive with the author's "beautiful peninsula" (bella penisola), which, once "squeezed in man's large electrical hand" (stretta nella vasta mano elettrica dell'uomo), will express "all its juice of wealth, beautiful orange so long promised to our thirst and finally conquered!" (Marinetti 1990g, 321; tutto il suo succo di ricchezza, bell'arancio da tanto tempo promesso alla nostra sete e finalmente conquistato!). While pointing to futurism's aggressive, consumeristic attitude toward land, food, and women, this image resonates with James Scott's description of high modernism as "the mystical combination of faith in scientific and technical progress and indisputable central power all aiming toward unlimited economic growth" (quoted in Armiero 2011, 36).

The first Italian oil wells were drilled in the 1860s around Parma. In the first decade of the twentieth century, others began to appear around Catania, Reggio Emilia, Frosinone, and Pescara (but also abroad in the Dahlak Archipelago in the Red Sea).[3] It thus seems reasonable to suggest that both their technological novelty and the implicit promise of prosperity for the nation played some role in Marinetti's imagining this miraculous "juice of wealth" buried under Italy's "skin." This was a nectar simultaneously able to free the untapped physical and intellectual potential of an exclusively male subject, speedily move him upward at the expense of everyone and everything else, and transform social, financial, and political dynamics. Despite the undeniable anthropocentric perspective and faith in man's ingenuity that informs Marinetti's words, it is also because of the vitalist agency of this juice and its inextricable relationship to human beings and the dynamics of a nascent capitalism that, as he optimistically puts it, "hunger and poverty have disappeared. The bitter social question has been annihilated. The financial question has been reduced just to production accounting. Freedom for all to make gold and mint gleaming coins!" (1990g, 321; la fame e l'indigenza scomparse. La amara questione sociale annientata. La questione finanziaria, ridotta alla semplice contabilità della produzione. Libertà a tutti di far dell'oro e di coniare monete lampanti).

However, as we learn from his almost coeval "La nuova religione-morale della velocità" (The new religion-morality of speed; 1990h), the actual place of origin and the production process of this miraculous substance remain unclear and ungraspable, ambivalently located somewhere between heaven's heights and earth's depths. As Marinetti writes, "Man stole electricity and fuels from space, to create new allies in the engines. . . . Wheels extract from the earth all the sleeping sounds of matter. . . . Gasoline is divine" (1990h, 130–37; L'uomo rubò l'elettricità dello spazio e i carburanti, per crearsi dei nuovi alleati nei motori. . . . Le ruote estraggono dalla terra tutti i rumori dormenti della materia. . . . La benzina è divina). In a way similar to Dario Bernazzoli, a Genoese artist who created advertisements for the Esso Oil Company throughout the

1930s and early 1940s, Marinetti too "elide[s] any reference to the messy geopolitical business of oil extraction, transport and refinement" (Pinkus 2012, 159) in his works. For both figures, oil and gasoline are ultimately associated with a metaphysical dimension that, while enigmatic, is also reassuringly and divinely pure and clean—always available to be safely tapped into by consumers.

Thus on the one hand, Marinetti's utopian, technologically infused but essentially nonscientific petro-genesis links electricity and fuels with speed, mobility, progress, wealth, and the supernatural dimension. On the other hand, petroleum is never perceived as a concrete, dirty, chthonic substance that will eventually play a role in producing electric energy itself. Most significantly, in the first part of "La guerra elettrica," one learns that electric power is generated by various natural sources, none of which includes oil.[4] First, there are the "diligent and furious storms" (tempeste diligenti e furiose) of the sea that provide "incessant movement to uncountable iron slabs in turn activating two million dynamos set along the beaches and one thousand blue-collar gulfs" (Marinetti 1990g, 319; moto incessante a innumerevoli zattere di ferro, che fanno funzionare due milioni di dinamo, disposte lungo le spiagge e in mille golfi operai). Second, there is the "energy of faraway winds and the rebellions of the sea" (energia dei venti lontani e le ribellioni del mare) that, in turn, are "transformed by man's genius in many millions of Kilowatts" (319; trasformate dal genio dell'uomo in molti milioni di Kilowatts). Finally, electricity production depends on collecting the "immanent atmospheric [and] telluric electricity" (elettricità atmosferica immanente) through "lightning rods and accumulators scattered on the rice fields and the gardens" (320; parafulmini e quei pali accumulatori sparsi all'infinito per le risaie e i giardini) onto a web of "metallic cables" (cavi metallic) that are able to carry the "double power of the Thyrrenian and Adriatic Seas up to the crest of the Appennines" (319; doppia forza del Tirreno e dell'Adriatico sale fino alla cresta degli Appennini). Qualifying Marinetti's fantastical ideas about obtaining unlimited energy in "La guerra elettrica" as early twentieth-century sci-fi is quite appropriate. Thus when Gerry Canavan, although referring to science fiction produced a couple of decades later, observes that oil's centrality to twentieth-century liberal capitalism is "proven precisely through the fantasy of its painless transcendence" (2014, 4), he might as well be referring to a context such as "La guerra elettrica," in which fuel and, by extension, oil are simultaneously visible and erased, ecstatically evoked but then also superseded by "something even more excessive and miraculous" (5) like electricity itself.

Marinetti's poetic solution in "La guerra elettrica" to return to the eternal natural forces of the sea, the wind, and the land to explain how electrical energy is generated reflects the time in which he was writing, when energy was still mostly generated from an increasing number of hydroelectric plants that aimed at eliminating Italy's dependence on foreign coal and other fossil fuels. As Marco Armiero observes, "In 1905 Italy occupied third place in the world production of hydroelectricity with seventy percent of its energy produced by water; between 1898 and 1914 the hydraulic power produced in Italy shifted from 40.000 kWh to 850.000 kWh. These numbers . . . represented millions of tons of concrete, iron and dirt, making artificial lakes, diverting rivers, changing the face of mountains" (2011, 34–35). From an ecocritical angle, we might also note a familiar sense of nostalgia and a lingering "presence of the past" within futurism. At the same time, Marinetti's bold plans to electrically stimulate plants' roots in order to enhance "the vegetable cells, the soil's nutritive principles" (le cellule vegetali [dei] principi nutritivi del suolo) and to irrigate and grow "large woods, immense forests" (grandi boschi, foreste immense) on the sides of Italian mountains, which would "whip the old cadaveric, tearful face of the ancient Queen of loves" (1990g, 321; sferzare la vecchia faccia cadaverica, solcata di lagrime, dell'antica Regina degli amori), could be perceived as an attempt to save a fundamentally sick and moribund nation—one that was already on a path toward environmental and cultural degradation. From this limited ecological perspective, Marinetti's projects to restore, both poetically and topologically, Italy as the garden of Europe and "love-room of the cosmopolitan world" (1990g, 324; love-room del mondo cosmopolita) does not seem such an illogical or unsustainable course of action. Once the country became a more rugged, wooded, and virile nation, it would be able to host not only a leaner, stronger, more agile population—renouncing heavy meals such as *pastasciutta*, as Marinetti and Colombo notoriously suggested in his *The Futurist Cookbook* (1989)—but also a natural, more vigorous landscape comparable with or even aesthetically and energetically superior to those belonging to competing European countries such as Austria and Germany.[5]

Thus while Marinetti's mystical infatuation with hydrocarbons, his creative energy production plans, and his bold landscaping operations are certainly appealing at the poetic level and may well signal futurism's authentic, nationalistic interest in the overall health of the Italian Peninsula, they are by no means indicative of a realistic concern for nature. These utopian futurist projects should rather be read as early, revealing examples of a broader rhetoric of regeneration and reclamation (bonifica) that, a few years later, would characterize fascist environmental discourses and practices that aimed to improve "not just 'external' nature but also 'internal' nature, people as well as places" (Armiero 2014a, 241–42; Ben-Ghiat 2001, 4). And ultimately, as history has taught us, neither petroleum, electricity, nor any other sources of energy would be enough to prevent these reclamation schemes from showing all their limitations and costs for both humans and their environment.[6]

If oil remains only partially represented in Marinetti's text, it becomes a much more visible protagonist just a few decades later in Maria Goretti's "Song of Petroleum." In this long futurist poem, oil becomes coextensive and entangled with the beautiful body of a black woman; it also has an enchanting voice with which it invites the gaze of man: "I sing! / with joy / with freedom / Listen listen man! / Look at me / I am beautiful / my body is glistening / a black dancer / a single scream / a long spurt screaming / toward the azureness" ([1941] 2009, 476–77). Echoing and at the same time "updating" Marinetti's original futurist aspiration to reshape and revitalize the decadent, feminized image of Italy and its territory, Goretti rejects the "ancient sweetness / of a poetic landscape" in both human and natural terms.

Mostly active during the 1930s and early 1940s, a period in which futurism and fascism often shared cultural visions and ideologies, and remembered especially for her *La donna e il futurismo* (Woman and futurism; 1941), a book that enthusiastically endorsed fascism's stereotypes of women as "hearth angels" and fertile mothers above everything else, Goretti's decisively antifeminist rhetoric and gender politics have much more in common with the misogynistic discourses of her male fellow futurists and fascists than with the emancipatory ones of earlier futurist women artists such as Valentine de Saint-Point and Rosa Rosà. Accordingly, in her poem, Goretti is not interested in growing forests and woods on the sides of the mountains of the *bel paese*. Rather, in a very different feminine landscape—one that strongly evokes colonial Ethiopia, given the period and the racially loaded reference to the "glistening / black dancer"—she inserts a "drilling tower / . . . / on the white slate of the sky / exact isosceles triangles of iron" that "builds harsh steel songs" and whose phallic "drill bores / sinks inseminates / bites crushes tears" into a screaming "violated earth" ("Down down the drill sinks / Deeper / the enemy earth says no nooo / yes yessss hisses the faith of men" [(1941) 2009, 476]).[7] Comparing this scene to the similarly violent one at the beginning of Marinetti's *Mafarka the Futurist*, in which a group of "beaten up and darting" (peste e guizzanti) African women is raped by male soldiers with their drill-like "blackish, smoky, organs" (1998a, 25; membri nerastri, affumicati), immediately reveals the overlap between the former scenario of resource exploitation and the latter one of "sexual colonization of a feminized land" (Cannamela 2019, 164). Goretti's fascist-inspired position in terms of gender relations and her justification of brutal and racist attitudes toward colonized people and lands also clearly emerges in these lines.

At the same time, and partially contradicting what is noted so far, in Goretti's vision, this explicit rape of the "enemy earth" is necessary to free the imprisoned, female-gendered petroleum from "the mountain upon [her]" so that she can ascend ("From the deepest / to the highest") to see, encounter, and finally possess a long sought-after "Light" that is a simultaneous source of orgasmic pleasure, power, and glory ("I see the Light now / and she will be mine mine mine . . .—My, my Light / Ah she leans / to kiss me / finally / all my body relaxes / quivers with pleasure" [(1941) 2009, 479]). While this light has been until now kissing an unaware "Man" who, as a consequence, was able to see "beauty," it has been denied to this feminized petroleum, buried and imprisoned by an overbearing weight above her.[8]

If there is no need to emphasize how this poem reutilizes and pays homage to stereotypical futurist themes and rhetorical patterns while enthusiastically singing the birth of a new energy source, it is tempting—despite Anna Nozzoli's observation that "the road of female narrative during the 'crisis' years of futurism . . . seems to move in the sign of . . . repeated antifeminist vocation" (quoted in Orban 1995, 69)—to individuate a distinctive hybrid feminine/material voice. This is to say, a voice that, after decades of relative marginality due to a "hindering mountain" that could be easily be named "Marinetti," wishes to finally be heard, see this empowering "Light" herself, and thus obtain a more active, productive (natural-cultural) role. In this sense, "Song of Petroleum" might be simultaneously considered as a site of subtle resistance to the sexist and far-from-emancipatory views of women informing both the traditionally male-centered futurist poetics and fascist ideology. This interpretation also indirectly corroborates Lucia Re's intuition about the different ambivalent ideologies in Goretti's prose and poetry, whereby the latter is seen as a "'ventriloquized' form of expression" with a "'secret' resistance-function . . . carrying out a poetic deconstruction of the fascist myth of femininity" (Re 1989, 271).

From this perspective, references to "the white page of the horizon," the "black gold," and in particular, to the "harsh embrace" with which this feminized petroleum "will kindle / [man's] sweet white flesh / with burning shivers" might be interpreted as allusions to the author's achievement of agency and the act of writing itself—as if oil was not only "living blood" but also a kind of ink ("liquid gold") able to leave its symbolic and material marks on different bodies, both natural and cultural, human and nonhuman. Thus somewhat paradoxically and counterintuitively, given its exploitative context, Goretti's poem suggests that only when a female writer appropriates and embodies the new, material, and semiotic force of oil and, vice versa, when this substance in turn contaminates and infects her—a joint cooperation and interference between entities belonging to different ontological realms—will both of them be liberated as they emerge from the dark recesses of (male) culture and the depths of nature, respectively. At the same time that the poem recognizes matter's agency, suggesting that the oil below the earth's crust desires to gush out, it also gives a sort of anthropocentric inevitability to the whole extraction process, for which it provides a convenient apology (i.e., since oil wishes to be liberated, all human efforts to do so are obviously justified).

Strange Emulsions: Oil Propaganda and Resistance in ENI's *Il Gatto Selvatico*

A brief historical overview may be helpful as we move on to decades that are even more crucial for the rise of the oil economy and continue to explore some of the cultural forms, expressions, and symbolic practices that promoted, contested, or showed ambivalence with regard to the modern diffusion of oil capitalism within the Italian context.

Although from the 1950s on, Italy was a proportionally minor player as a producer in the global oil industry, it was also one of the first countries that switched from coal to hydrocarbons as an energy source during the nineteenth- and twentieth-century industrialization process. Given its favorable geographic position along the Mediterranean "oil road" halfway between northern Europe and Africa, it hosted (and still hosts) some of the largest petrochemical centers and refineries in Europe and the world. In particular, from the early 1950s through the 1970s, petrochemical plants, resulting from the combination of economic calculations with the country's topography and high population density, were often built by both Italian companies (such as ENI [Ente Nazionale Idrocarburi] and Montedison) and Anglo-American ones (i.e., Exxon, Gulf, BP), often in the proximity of urban centers and in areas of distinctive natural beauty. In his article "Petrochemical Modernity in Sicily,"

Salvatore Adorno observes that "by the 1970s, petrochemical plants with names such as Raisom, Sincat, and Celene, owned variously by Esso, Montedison, and other companies, spread across some 2.700 hectares of coastal Sicily [between the cities of Augusta and Siracusa]" (2010, 180). Serenella Iovino, in her *Ecocriticism and Italy*, poignantly reminds us that right next to the city of Venice, famously despised because of its association with "pastism" by Marinetti and fellow futurists, there is a fundamentally unknown "anti-Venice": "the Montedison petrochemical factory of Porto Marghera, just a few miles from San Marco Square" (2016a, 48).[9]

A quick glance at the locations of some of the other main plants that were set up in these decades—near Ferrara and Mantova (in the region of Lombardy), Priolo and Gela (in Sicily), Ravenna (in Emilia Romagna), Leghorn and Rosignano (in Tuscany), Brindisi and Manfredonia (in Puglia), and Assemini, Sarroch, and Porto Torres (in Sardinia)—confirms their ubiquity across the national territory and retraces a fundamental map of Italy's petrochemical development from the years of the "economic miracle" on.[10] It also hints (especially, but not exclusively, with regard to the South and the islands) at the neocolonial considerations that went into locating high-risk, polluting industrial plants in areas of poverty and low levels of education. Even though places like Mantova, Venice, Ferrara, and Ravenna may now be considered affluent urban centers, their industrial peripheries—especially in the period under consideration, when there was a substantial internal migration of Southerners moving north—were mostly populated by people who saw in the newly established plants nothing more than a good opportunity to find work.[11] Additionally, as Paola Bonifazio suggests, ENI's plant distribution should be also viewed in light of the so-called Southern Question. This question broadly refers to the discourse about "the economic, social, and cultural divide between the modernized and industrial North and the pre-modern and rural South of Italy [and] . . . was essentialized and racialized by both Northern and Southern intellectuals in their writings ever since Unification" (2014, 331n6). Rather than suggesting that the South and the North were both contributing to Italy's modernization process, thus shifting the focus "from the Southern Question to issues of productivity" (341), ENI's films possessed an essentially neocolonial attitude toward the South and continued "to identify the South with the other to be rescued, the poor to be fed, and the worker to be educated, all by a charitable and compassionate North" (341).

ENI evolved in 1953 from the older Agip (Azienda Generale Italiana Petroli; General Italian Oil Company) with the aim to locate new oil fields and acquire, transport, distribute, and commercialize oil and its derivatives. Today it is a multinational company employing almost ninety thousand people all around the world. ENI was *the* propeller of the oil economy in Italy and one of the major players in its transformation into a modern, industrialized country during the "economic boom" that followed the devastation of World War II. The short promotional video entitled *ENI e il Boom* (ENI, n.d.) is historically significant to the fluidity and articulation of the petro-landscape under examination, since it points to ENI's multifaceted efforts to influence the beliefs, lifestyle, and choices of a rapidly transforming nation.[12] This corporate advertisement perfectly illustrates the wide range of possibilities that Italians (men in particular) would enjoy in the late fifties and sixties thanks to petroleum's agency and, of course, to ENI's crucial role in its discovery, extraction, and distribution. The fast-paced sequence of frames following the initial image of a gushing well informs us that oil creates not only new factory jobs but also the chance to enjoy a series of leisure activities, most of which involve automobiles and the freedom of movement they afford. Accordingly, footage of a man filling up at a gas station is followed by an aerial shot of a motor-racing circuit and then by a parking lot with three almost identical Fiat 500s in the foreground. The message is clear: even if you own a tiny economy car, once it is fueled up, you can be successful and have fun at the wheel just like a professional driver. Expectedly, in the next frame, a man chivalrously opens the door of his new Fiat 500 to a pretty young woman while holding a bouquet of flowers in his other hand. Owning a car also allows families to go on vacation or to have picnics with friends. This commercial ends with an image of the company's main logo (still in use today), the "six-legged dog" that, according to the official interpretation, represents "the sum of a car's four wheels and the two legs of its driver"—a perfect emblem of a posthuman assemblage, or even a new form of species companionship.[13]

This is not the place to review the history and achievements of a corporation that strives to project an equitable and respectful "Italian way" of doing business, which is allegedly distinct from the more aggressive and rapacious behavior of other leading, foreign-owned companies in the global petroleum industry. Georgiana Banita perfectly captures ENI's controversial stance toward oil-producing countries when she observes that the company's position was "vociferously anti-imperialist and sympathetic to [these countries'] struggles for independence, yet also calculating and self-serving" (2014, 149). I also do not intend to address the still mysterious, unsolved death of its controversial founder, Enrico Mattei (1906–1962), that has obsessed generations of Italians or to investigate the related commingling of economic power financed by petrochemical lobbies with Italian history, international politics, and organized crime. A large number of investigative journalistic reports and fictional works have already tackled these issues.

Among these cultural artifacts is the complex unfinished novel by Pier Paolo Pasolini, *Petrolio* (Petroleum; [1972] 1992), which allegorically voices the writer's indignation toward the dark, consumerist turn that Italian society was taking at the time. As it scandalously deals with "sex, politics and ENI" (Benedetti 1995, 11), *Petrolio* assigns responsibility to an indiscriminate neocapitalist and neofascist development that managed to erase all the cultural formations and expressions resisting the advancement of a new consumerist totalitarianism. *Petrolio* is thus an eminently political novel that imaginatively illustrates some of the wide-ranging consequences of Italy's transformation from a rural to an industrial country or, as Rebecca West acutely observed, serves as an example of *écriture corporelle* (corporeal writing) that searches for a protoecological "circularity between subject and object, the internal and the external," thus bypassing those "conceptual binarisms that . . . keep conditioning our Western (and patriarchal) thought" (quoted in Benedetti 1995, 44). In the words of Karen Pinkus, it is also "an alchemical, novelistic, organicist work that explores the profound ties between fossil fuels and narrative [and that even though it was] written before a widespread consciousness of climate change, seems impressively to anticipate, and somehow to correspond intriguingly with, the chaotic temporality of the Anthropocene" (2014).[14]

Rather than underline Pasolini's ecocritical relevance in the subfield of petrofiction, in what follows, I address a selection of the lesser-known and critically unconsidered promotional narratives originally commissioned by Enrico Mattei. These pieces were composed by several Italian intellectuals, writers, and filmmakers (many of whom were Pasolini's contemporaries) and appeared in the pages of ENI's first official company publication.[15] Born in 1955 out of the collaboration among founder Enrico Mattei, poet Attilio Bertolucci, and graphic designer Mino Maccari, with a title that means "wildcat" in English (a technical oil field term that indicates an exploration well), *Il Gatto Selvatico* was published until 1964, and one might think of it as a perfect emulsion of petroleum and storytelling. It was informed both by its editors' awareness of the power of narratives to shape the world and by the belief that industry and culture—that is, technological and humanistic discourses—were not mutually exclusive but could actually contribute to creating a more human, positive, and reassuring kind of industry along the lines of an enlightened neocapitalism.[16] As Enrico Mattei himself put it, "[Our magazine] must function as a meeting-point for all those who belong to the great family of the ENI Group.... The magazine we shall make must be the same for everyone—that is to say, it should be accessible in a democratic way for both the President of the Italian Republic and the farthest away of our drillers, even outside of Italy" (quoted in Belloni 2014, 6).

While targeting the largest and most heterogeneous audience possible, *Il Gatto Selvatico* purported to educate and inform its readers of the many exciting changes that an ENI-centered fossil fuel economy would bring to virtually every aspect of life, including geographic as well as anthropological landscapes. Of course, a parallel objective was also to help ENI promote a view of oil as fundamental for living a modern existence and, at the same time, to conveniently shift its audience's attention away from any criticism or questioning of the socioenvironmental costs associated with oil extraction and processing and its widespread applications. Accordingly, a typical issue of the magazine consisted of brief, descriptive cultural pieces combining a heterogeneous range of topics (technology, science, urbanism, architecture, art history, cinema, theater, literature, sport, fashion, leisure, tourism) written by both well-established and emerging contributors. Looking at some of the literary sections through their years of publication is especially revealing. Along with some minor figures, one finds many of the writers that would shape Italian letters and intellectual life in the twentieth century, including (in quasi-chronological order of publication) Giorgio Caproni, Giovanni Comisso, Goffredo Parise, Giuseppe Dessì, Enrico Pea, Carlo Emilio Gadda, Giorgio Manganelli, Giuseppe Berto, Gianna Manzini, Raffaele La Capria, Giorgio Bassani, Mario Soldati, Luigi Davì, Alfonso Gatto, Anna Banti, Alberto Bevilacqua, Carlo Cassola, Italo Calvino, Leonardo Sciascia, and Natalia Ginzburg.

Il Gatto Selvatico was not just an industrial publication but also a laboratory that believed and invested in culture, welcoming and promoting different writers and providing a colorful and optimistic portrait of Italy in the fifties and sixties. Yet no specific ecocritical attention has been given to the role that, by requesting some contributors to find creative ways to celebrate the glories of the "dark gold," the magazine had in "establish[ing] the conditions necessary for a dynamic oil-economy" (Barrett and Worden 2014, xxvii). Thus the publication exploited the marriage between humanistic culture and technology for ENI's self-serving, corporative purposes. A closer look at a representative sample of these short stories and at the substance that generates and fuels the narratives should further clarify this claim while revealing how such a celebration also presented ambiguities and resistances.

As it narrates the experiences of a writer who, at the suggestion of a relative who owns a car, decides to stay at the "Motèl" in Rome with his family, Francesco Squarcia's simple and straightforward "Un pedone al Motèl" (A walker at the motel; 1959) is a brutally explicit example of what I would call "promotional petrofiction." The text wastes no time in making the point that the Motèl—this new kind of lodge for car drivers that ENI's predecessor, Agip, imported from America and started building along major roads and freeways all over Italy[17]—is the modern, fashionable object of desire and sensible choice for travelers and tourists. Why, the story posits, should one stay in a nineteenth-century hotel decorated with ancient furniture, smelling of wallpaper adhesive and catering to "old bejeweled ladies with too much makeup" (1959, 15; vecchie ingioiellate e ritinte) when there is the option of lodging in an odorless, bright, functional, American-looking place? This, after all, is a place that mostly hosts a classy, foreign clientele, where

> customers are fluent and young, even though they are sixty years old, and the light-colored metal elevators ascend effortlessly. One feels very far away from a city so full of history. . . . Here we are outside of history; we start from zero.

> La gente è spedita e giovane, anche se ha sessant'anni, e gli ascensori di metallo chiaro vanno su con leggerezza. Pare di essere lontanissimi da una città così zeppa di storia. . . . Qui siamo fuori della storia; si comincia da zero. (15)

The answer is simple. Indeed, with a rhetoric that strongly resembles the Nietzschean language of the futurists just a few decades earlier, the author's implicit invitation is that Italians should forget the symbolic heaviness and existential lowlands of their rural and poverty-stricken past. From this new architectural tabula rasa, they should effortlessly and dynamically move on (and upward, in social and economic terms) along with progress, either in an elevator or, even better, in an automobile, since walking is for pedestrian losers.[18] In fact, as the narrator states, "We received perplexed glances from the concierge when they realized that we were beginning to walk" (Squarcia 1959, 16; Dal bureau ci guardarono un po' perplessi vedendo che ci avviavamo a piedi).

Ironically, it is Rome's classical historic center—with its urban sounds, trams, crowds, and "violent life" (vita violenta)—that emerges as a more appropriate, dynamic site of modernity, while the new Motèl's peripheral location is associated with the quiet of an Arcadian countryside located far, but not too far, from the city. This is a place where "one could hear the crickets singing in the fields" (Squarcia 1959, 17; si sentivano per i campi cantare i grilli) and one that fosters pleasant, tranquil conversations. In other words, it is a site that cars approach "in silence, as if they were tame and hungry animals" (17; in silenzio, come animali docili e affamati), and where every operation is carried out in a "regular, soft, mysterious rhythm" (con un ritmo regolato, soffice, misterioso) while the sky reverberates "a milky dim light as if at dawn" (17; un chiarore latteo come di alba). But this is only an apparent contradiction, since the message *Il Gatto Selvatico* wants to convey to all the aspiring owners of automobiles and future Motèl guests in the nation is that having a car, driving it, and filling it as often as possible with Agip gas are the most natural and nondisruptive things to do—acts that are in total harmony with the rhythm of nature and the universe. In the context of a Roman countryside, the Motèl guests ideally substitute shepherds, and automobiles take the place of "tame and hungry" sheep and goats—instead of grass, they need to be fed with gasoline. Rather than functioning "culturally as an alternative to the urban technological practices of modern and postmodern fossil-fueled capitalism" (Sullivan 2015), the classical trope of the pastoral is here diabolically absorbed and put to the service of "fossil-fueled capitalism" in a dynamic that could be defined by the neologism *petropastoral*.[19]

Squarcia is not the first to have adopted this sort of rhetoric and form. Straying momentarily away from *Il Gatto Selvatico* and going back in time a couple of decades, Massimo Bontempelli's short novel *"522": Racconto di una giornata* ("522": Tale of a day; 1991) could be considered one of the earliest examples of the Italian petropastoral. Bontempelli's work is an explicitly promotional text that another fully oil-dependent corporation, Fiat Automobiles, commissioned to boost the launch of its 522 compact-car model. Building on yet also distancing himself from the futurists' infatuation with machines, Bontempelli's main (corporate-directed) objective is to domesticate this new, intimidating technological object and make it appealing for buyers. He accomplishes this by first reassuring readers of the car's symbiotic and harmonious relationship with nature and then by anthropomorphizing it, thus providing it (or, from now on, "her") with familiar human feelings and psychological reactions.[20] Accordingly, when the 522 rolls out of the Lingotto factory in Turin on a bright spring morning, she is welcomed by the natural world as if she were a newborn sentient being who lives, feels, and even feeds, anticipating the animallike cars mentioned by Squarcia:

> The first time that 522 came out and went around the world was a morning in May.... At the sides of the road, two rows of poplar trees sweetly waved their pale trunks and crowns, fanning the road from above, and the air lovingly descended on the beautiful asphalt strip, the hood, and the shining skin of the radiator. 522 is now speeding up, pleased to live and discover the land. There is a lot of world made of green expanses up to the ethereal curve of the horizon, right and left, even beyond the road and the two rows of poplar trees.... There was a gas pump and 522 stopped and drank.
>
> La prima volta che 522 uscí dal chiuso e andò per il mondo, era una mattina di maggio.... Ai due lati due file di pioppi ondulavano dolcemente i pallidi tronchi e le chiome: queste dall'alto cosí fanno vento alla strada, l'aria scende con amore sul bel nastro d'asfalto, e intorno al cofano bruno e contro l'epidermide fulgida del radiatore. 522 corre, mossa dal piacere di vivere e scoprire la terra. Anche oltre la strada e le due file di pioppi c'è molto mondo, a destra e a sinistra, fatto di verdi distese fino al circolo etereo dell'orizzonte.... Lí c'era una colonna della benzina, e 522 a buon conto vi si fermò e abbeverò. (Bontempelli 1991, 3–4, 7)

The idea of an innocuous and optimistic merging of technology, humans, and nature is even more powerfully reiterated at the conclusion of this story, which imaginatively evokes the Edenic and utopian topography of a vibrant, joyful, and free country that is only waiting to be discovered by the 522 and her driver:

> She knows that many more days and outings are waiting for her, in which she'll feel free, powerful and will endlessly breathe in the wonderful Italian air; she'll speed carefree on the clear asphalt, along the three azure seas, around the joyful lakes, between palpitating forests, toward the gleaming mountains; in bright days and soft nights; under the most exciting sun and the most shining constellations; among the most fervid human, vegetable and animal nature: the sun and the stars and the water and the air and the lives [all living] in the happy gardens of the world.
>
> Sa che molte giornate ancora e molti cammini la aspettano, di libertà e di potenza, di respiri infiniti tra l'aria bellissima d'Italia; sugli asfalti limpidi abbandonato correre, lungo i tre mari azzurri, in giro ai laghi allegri, tra i boschi che palpitano, verso le montagne che scintillano; nei giorni accesi e nelle notti morbide; sotto il sole piú eccitante e

> le costellazioni piú lucide; in mezzo alla natura d'uomini e di piante e d'animali piú fervida: il sole e le stelle e l'acqua e l'aria e le vite del giardino felice del mondo. (83–84)

Finally, the fact that those same "happy gardens of the world" (84) that, in Bontempelli's vision, constitute the ideal playground of the 522 and Fiat in the thirties are the same ones surveyed by ENI's engineers in the fifties as they scouted for plant locations conveys the close historical connection between these two major Italian companies.[21]

Closing this parenthesis on Bontempelli and returning to *Il Gatto Selvatico*, Luigi Daví's story "I centauri" (The centaurs; 1959) seems a natural follow-up to the form of the petropastoral. As Greek mythology reminds us, centaurs—hybrid, liminal beings who were half human and half horse—were among the inhabitants of the literary region of Arcadia, where they occasionally broke the peace, innocence, and contentment of that bucolic landscape because of their lustful, unruly, and savage attitudes.[22] Half men and half motorbikes, Daví's modern centaurs come out of an urban environment, and yet they follow a behavioral pattern similar to that of their classical predecessors. They venture in groups out of the city into the rural province in search of weekend fun, dance, conquest, and romance with willing local girls, whom they view as goods to be enjoyed and then discarded: "Every Sunday there was a new town to discover.... We started to ride in a single line, then followed and passed each other until we reached our destination" (1959, 24; Ogni domenica era un paese nuovo da scoprire.... Poi via in fila indiana ad inseguirsi e superarsi finché si fosse giunti). Provided that, in a recurrent image of animalization of the technological object, their two-wheeled vehicles will not be mistaken for cows by the naive local parking attendants ("As long as our bikes won't be parked in a stable and milked like cows" [24; Purché non ce le mettano in stalla con l'idea di mungerle]), these bikers can look forward to more empowering escapades, male bonding, and adventures. They will hop on their bikes and, continuing to burn "super" (i.e., the best quality of gasoline), achieve both an improved mechanical performance for their rides and a better sexual one for themselves:

> —"There are no girls here. Why stay?"—said Ezio.—"There is a party in Bussoleno: who wants to go?"—another excitedly added... "Follow me, kids!"—shouted Leo, who had rigged his engine and burned super. They entered the road and loudly stepped on the gas.

—"Tanto qui non ci sono piú ragazze: che ci restiamo a fare?"—ribadì Ezio.—"A Bussoleno c'è veglionissimo: chi è d'accordo?"—rilanciò uno più eccitato.... "Statemi dietro, pulcini!"—esclamò Leo che aveva truccato il motore e bruciava "super." Sboccarono sullo stradone e diedero tutto gas, fragorosi. (27)

How could this scenario of conquest, freedom, and mastery of both roads and women not be an appealing and desirable one for a whole generation of young Italian men who were growing up in a mostly straight and unquestionably patriarchal country?[23]

While the oil-rich message in Squarcia's and Davì's contributions is hard to miss, Giuseppe Dessì's "Il fidanzato" (The fiancée; 1955), which deals with the potentially serious consequences of a kids' prank in the Sardinian village of Norbio, is a subtler form of promotional literature. However, the recurring presence and central narrative role of an old oil lamp, together with the first-person narrator's insistent reflections on the lack of electricity in the village and, therefore, of primitive, shadowy nights filled with fear and a doubt-inducing darkness, make clear that this story is also an implicit ode to the enlightenments brought by technological progress and by new, liberating forms of petrol-generated energy. As the narrator repeatedly observes, the presence of electricity would have defused the drama and excesses caused by the prank—a male-looking leather puppet placed in a young woman's bed—and eliminated subsequent violent reactions, fears, and lingering suspicions:

> Now I think that everything would have remained just a big joke, if only electric light had been there; but the line arrived at Norbio only a few years later. . . . At that time (perhaps because of the lack of electric light) life, which was so peaceful, tranquil and stable during the day, became full of doubts, incertitudes and religious fears at night.
>
> Ora penso che tutto sarebbe rimasto entro i limiti di una grossa burla, se ci fosse stata la luce elettrica; ma solo alcuni anni dopo la linea arrivò fino a Norbio. . . . A quel tempo (forse a causa della mancanza di luce elettrica) la vita, così pacifica, tranquilla e stabile di giorno, si rivelava invece di notte piena di dubbi, di incertezze e di religioso timore. (Dessì 1955, 15)

In other words, as Marinetti envisioned, electricity is a sign of progress and modernity, and unless one wishes to linger in the dark ages and ignorance of a primordial past, it is an advancement that, like the energy petrochemical companies will generate and distribute, ought to be trusted and welcomed on the island. In addition, the lurking paternalism of the author also informs us that resisting or delaying this change may even threaten the whole integrity of the family, since women are in danger of finding unknown men in their beds.

Problematizing Oil's Celebration

The last works I wish to consider in this vein are Carlo Emilio Gadda's "Il pozzo n. 14" (The well n. 14; 1960) and Leonardo Sciascia's "Gela: Realtà e condizione umana" (Gela: Reality and human condition; 1964) and the related 1964 documentary to which he contributed with Giuseppe Ferrara, *Gela antica e nuova* a.k.a. *I miti e il petrolio* (Gela ancient and new a.k.a. The myths and petroleum; 1964). These works, despite their propagandistic context and apparent proindustry stance, cannot be considered fully celebratory. On the contrary, they convey a troubling picture of the existential costs and sociocultural implications of technological progress in general (Gadda) and industrial petrochemical production in particular (Sciascia). How could these authors get away with anticelebratory texts? The ubiquity and confidence of petroculture and, in particular, the farsightedness of Mattei and the editors of *Il Gatto Selvatico*—who had nothing to fear and more to gain from the inclusion of these literary figures—could well be a determining factor. Additionally, along with Stephanie LeMenager, we may also recall that printed language itself is ultimately "petroleum based" and, as it provides another indirect evidence of oil's pervasiveness, it may help explain ENI's only apparently counterproductive editorial policies (see LeMenager 2012, 63–64).

Before appearing in *Il Gatto Selvatico* in 1960, Gadda's short story was originally published in the newspaper *La Gazzetta del popolo* (1934) and included later in his book *Le meraviglie d'Italia* (The wonders of Italy; 1964). This volume collected writings on the world of technology inspired by his travels and professional experiences as a hydraulic engineer from the twenties to the forties. He allegedly aimed to celebrate both the achievements of Italian industry in Italy and abroad and the bright future the nation had ahead of itself. However, the presence of ambivalent, ironic attitudes and tensions in Gadda's work indicates that his praise of technology is also often accompanied by criticism and resistance, his playful enthusiasm by skepticism, and his faith in progress by a profound sense of anguish. His "poetics of polarity," alternating between "lyricism and satire, subjectivity and mimesis, historical awareness and reactionary blindness" (Sbragia, quoted in Past and Amberson 2016, 28), shows a simultaneous awareness of the extent of both human accomplishments and "the tragedies and indignities of anthropic time" (Past and Amberson 2016, 70). What I am suggesting, in short, is that "Il pozzo n. 14" is another example of this sort of fluctuation and contradiction.

Set in the village of Carlingen in French Lorraine near the border of Germany, the story deals with Gadda's own work experience there as a technical manager of a chemical plant producing ammonia, revealing the existential demands, difficulties, and challenges (mechanical, geographical, emotional) connected to this "new technological audacity" (Gadda 1964, 43; nuova audacia tecnologica). Given its industrial topic and the fact that ammonia sulfate, used as a fertilizer, is also a standard by-product of petrochemical plants, it is thus still relevant in our discourse, even though it does not deal specifically with oil or oil extraction.

Any risk of mistaking this short story for a simple, optimistic celebration of human managerial, entrepreneurial, and technical skills is immediately dispelled from the beginning. As it describes the rural surroundings of the village and the place where the protagonist will reside during his stay, the story relates and contrasts the initial, playful natural image of the "square dunghills overflowing with an abundance of horseshit" (Gadda 1964, 40; concimaie quadrate, ricolme di uno strabocchevole e dovizioso letame cavallino) emitting "warm vapors" (40; caldi fumi) with the conclusive picture of an anguish-ridden, hopeless, sooty industrial landscape in times of war amid "woods that, through the smoggy haze, beyond the smokestacks and the towers, I [the protagonist] could see blackening in a blind horizon" (47; selve che ora, tra la foschia de'fumi, vedevo annerarsi di là dalle ciminiere e dalle torri, nel cieco orizzonte). Gadda's literary universe, Past and Amberson remind us, is often characterized by a "vision of dynamic materiality where beings and objects are defined (and redefined) by their relation to a multifaceted reality" (2016, 68–69). Here, the warm, innocuous, animal-produced vapors—which used to "exhale a prophecy of spring" (Gadda 1964, 40; esalando un vaticinio di primavera) and later transform into sinister, smoggy plumes of smoke that obscure the woods and any hope for the future—are likely to fall within such a dynamic vision. While the attention shifts from the scatological to the industrial, from the organic to the inorganic, the protagonist remains sensorially and materially enmeshed with the fundamentally smelly and shitty world around him.

Similar to an organic body, the chemical plant also requires constant care and effort to make it function properly. Recalling Bennett's notion of "vital materiality" (2010, 6), it seems to breathe on its own ("Some misbegotten wash-pumps... had worked all night to blow air into the gas" [Gadda 1964, 41; Certe malnate pompe di lavaggio... avevan lavorato tutta la notte ad insufflare aria nel gas]) and even evacuates:

> A quick emptying relieved me from undesired responsibilities... we saw it blowing out all its evil smell... the batch of coke-ovens opened some mouths to unload the Schlacken, the incandescent scorias.
>
> Un pronto vuotamento mi sollevò da indesiderate responsabilità... lo riducemmo a soffiar fuori tutta la sua puzza malvagia... la batteria dei forni a coke apriva taluna delle sue bocche per scaricare le Schlacken, le incadescenti scorie. (Gadda 1964, 41)

Just like an organic entity may give an initial appearance of solidity and stability and only later reveal all its faults and vulnerability, the impression of thickness and "powerful strength" (potente robustezza) of the devices, pipes, valves, and instruments of the chemical plant does not correspond to reality but is rather the product of a "delirium caused by a strong fever" (43; delirio d'una gran febbre).

Symptomatic of Gadda's ambivalence toward this industrial site, as well as of the limitations it imposes on the development of human subjectivity, are the protagonist's reflections on the kind of uncanny effect the plant and similar cathedrals in the desert have on him. As he puts it, this eventually becomes an isolated and isolating nonplace characterized by "a climate without past and intimacy, where the foreigner meets and does not greet another foreigner" (Gadda 1964, 47; un clima senza passato e senza intimità, dove lo straniero incontra e non saluta lo straniero) and where it is ultimately difficult to establish relations, to be interconnected and part of a community, or to put it differently, to feel that "all our lives are 'a symbiosis with the universe'" (Past and Amberson 2016, 69).[24]

On the contrary, in this techno-industrial context—an alienating site that "wasn't either France or Germany" (Gadda 1964, 47; non era Francia, non era

Germania) or, even less, Italy—the protagonist eventually withdraws and falls back into a nostalgically infantile universe. In other words, he withdraws into the narrowly nationalistic and parochial borders of an anguished "demarcated subject" (Past and Amberson 2016, 69), one who struggles and yet is fundamentally unable to establish a real connection with a larger, multiethnic exterior world:

> Oh! I did not feel European or a supporter of Europeanism . . . but I dreamt of Spoleto and Fiesole . . . and my mind, as I became a child again, momentarily re-created the amazed clarity of a fairy tale.
>
> Oh! Non europeo né europeista io mi sentivo . . . ma sognavo di Spoleto e di Fiesole . . . e la mia mente, ritornato fanciullo, ricreava un attimo la stupita limpidità della favola. (Gadda 1964, 47)

A strong autobiographical element also characterizes Leonardo Sciascia's nonfictional short essay "Gela: Realtà e condizione umana," in which the writer witnesses the various transformations that an aggressive oil-based industrialization brings to this historically poor village, located on the southern coast of Sicily. Similar to Gadda's, Sciascia's text is far from a mere eulogy of progress and industry, and he begins by pointing out how a fundamental poverty has affected this geographic area throughout its history, from its agricultural and sulfur-mining past, to its brief period of archeological fame, to its current, oil-rich present. As Sciascia puts it, "Personally, I think I never had before, as I had then in Gela, a harsher revelation of Sicilian poverty, of its misery" (1964, 18; Personalmente, credo di non aver avuto mai, come allora a Gela una più cruda rivelazione della povertà siciliana, della miseria). Sciascia was ahead of his time, as he was able to see that the abrupt switch from an agricultural way of life to an innovative yet also profoundly disruptive industrial infrastructure was not an absolute guarantee of socioeconomic wealth, safety, and hope. As a matter of fact, he observes that the town, after becoming "the land of oil" (20; la terra del petrolio),

> suddenly appeared to be poorer, the poor felt even poorer: commerce increased, but all the things crowding the shops seemed to push back the population's life to an even more obscure and distant zone. And there was diffidence: foreigners and Northerners believed, perhaps, that misery was more a vocation than a condition; and from their

poverty the inhabitants of Gela looked at those well-dressed and well-shaved men, at their families, their well-fed children, and considered them emblems of a new and different plunder launched on a land that had already suffered many depredations through the centuries.

> il paese sembrò anzi farsi di colpo più povero, i poveri si sentirono più poveri: saliva il commercio ma le cose di cui i negozi si infittivano sembravano respingere la vita della popolazione in una zona ancora più oscura e lontana. E c'era diffidenza: stranieri e uomini del nord ritenevano, forse, che la miseria fosse una vocazione più che una condizione; e dalla povertà i gelesi guardavano quegli uomini ben vestiti e ben rasati, le loro famiglie, i loro bambini ben nutriti, come i cursori di una nuova e diversa depredazione lanciata su una terra che già tante ne aveva subite nei secoli. (20)

It is only at the conclusion of his short essay—in what seems a last-minute realization that he is actually writing for the official journal of ENI—that Sciascia attempts a belated and not very convincing balancing act as he more optimistically observes how many people have been able to leave their primitive dwellings (*i catoi*), how incomes are rising, how access to consumer goods is wider, and finally, how

> there is always a festive atmosphere in town, similar to the one that, on Saturday evening, characterized some sulfur-producing villages, in the golden-age of the sulfur mines, but without that component of desperation and anguish ... without that glimmer of death accompanying the euphoria caused by wine.
>
> nel paese è sempre un'aria di festa: ch'era in certi paesi zolfatari, nei tempi d'oro delle zolfare, il sabato sera; ma senza quella componente di disperazione e d'angoscia ... senza quel baluginare di morte che era nell'ebbrezza del vino. (20)

However, if it is true that oil in Sicily is just the new sulfur, taking its place as a source of employment and income, this conclusive paragraph expresses a cautious hope for a better future while also betraying a somber premonition that a history of death and desperation may well repeat itself.

Thirty years after the publication of Sciascia's essay, Vincenzo Consolo's apocalyptic description of an environmentally and culturally degraded Gela in his Sicilian travel novel *L'olivo e l'olivastro* (The olive tree and the oleaster; 1994) cannot but posthumously confirm the skeptical lucidity of Sciascia's earlier diagnosis. Consolo's vision is dramatically split between his awareness of the necessity for some kind of socioeconomic development of the area and his serious concerns regarding the neocolonial implications of the industrialization process of the Italian South and the risks it poses to local culture and traditions:

> [It is] this obscene fetus of power and progress.... This new and unexpected Gela that separates the technicians, the geologers and the accountants coming from Metanopoli... and the locals involved in building abusive and wild constructions, brick houses and faulty rebars amidst the mud and trash of stuck-together neighborhoods, unnamed roads. This is the new Gela, with its sea fat with oils, its concrete breakwaters, its harbor dotted with stranded, leaning ships, looking like islands made of rust, plastics and rats... this is the Gela where every memory and sense has been lost, where minds freeze and aphasia reigns, where the languages spoken are only the disgraceful ones of syringes and knives, loud exhausts and TNT.

> [È] questo feto osceno del potere e del progresso.... Nacque la Gela repentina e nuova della separazione tra i tecnici, i geologi e i contabili giunti da Metanopoli... e gli indigeni dell'edilizia selvaggia e abusive, delle case di mattoni e tondini lebbrosi in mezzo al fango e all'immondizia di quartieri incatastati, di strade innominate, la Gela dal mare grasso d'oli, dai frangiflutti di cemento, dal porto di navi incagliate nei fondali, inclinate sopra un fianco, isole di ruggini, di plastiche e di ratti... la Gela della perdita di ogni memoria e senso, del gelo della mente e dell'afasia, del linguaggio turpe della siringa e del coltello, della marmitta fragorosa e del tritolo. (1994, 77–79)

A similar ambivalent and conflictual position toward industrial progress characterizes the coeval (and homonymous) documentary based on Sciascia's text: *Gela antica e nuova* a.k.a. *I miti e il petrolio*.[25] This short film was among the many that ENI commissioned over the years, and it is in the tradition of Joris Ivens and Paolo Taviani's *L'Italia non è un paese povero* (Italy is not a poor country; 1960); Guido Arata's *Petrolio nelle dune* (Oil in the dunes; 1967); Gilbert Bovay's *La valle delle balene* (The valley of the whales; 1965), *Gli uomini del petrolio* (Oilmen; 1965), and *Africa, nascita di un continente* (Africa, the birth of a continent; 1967); Bernardo Bertolucci's *La via del petrolio* (The way of petroleum; 1967); Gillo Pontecorvo's *Una storia per l'energia* (A story for energy; 1984); and Folco Quilici's *Uomo, ambiente, energia* (Man, environment, energy; 1988). During the 1960s in particular, ENI hired already (or soon to be) celebrated filmmakers to record and divulge its discoveries and technological achievements in Italy and abroad, showing its importance as an indispensable engine of local and global socioeconomic growth.[26]

Gela antica e nuova was directed by Giuseppe Ferrara (1932–2016), who, following his neorealist roots, would become one of the major representatives of twentieth-century Italian political and socially engaged cinema.[27] Scripted by Sciascia and dedicated to Mattei's memory, the film aimed at providing a visual narrative of the building of one of ENI's largest petrochemical plants in Italy after an oil field was discovered near the town. Anticipating the aesthetic stance that has been said to characterize Bertolucci's 1967 film, one could argue that this work too "breaks with both corporate documentary style and neorealism itself by infusing the story of oil with unusually poetic fervor, paving the way for the historicist lyricism of Pasolini and Francesco Rosi" (Banita 2014, 147–48, 163). And similar to the works of the latter figures, Ferrara's film undoubtedly manages to merge corporate propaganda and poetry.

In this sense, the opening sequence is exemplary. With a melancholic soundtrack playing in the background, it combines footage of a still-pristine, arid, Southern Mediterranean marine landscape, with its resilient low vegetation and prickly pear trees overlooking the long streak of yellow sand and strikingly blue sea in the Gulf of Gela. The setting is accompanied by a voice-over recitation of the first verses of "A un poeta nemico" (To a hostile poet; 1956) by Sicilian-born and Nobel Prize–winning Salvatore Quasimodo (1901–1968). In this poem, Quasimodo nostalgically evokes the "the straw-colored sands of Gela" (1966, 45; sabbia di Gela colore della paglia) where he used to lie down "as a child, by the ancient sea / of Greece, clenching many dreams in my fists, / on my breast" (45; fanciullo in riva al mare / antico di Grecia con molti sogni nei pugni / stretti e nel petto).[28] With the camera moving from the beach to show the archeological wealth of the nearby ruins of the ancient Gela, the recitation of the poem ends on a verse centered on Aeschylus's exile and his peculiar death, which happened right on this shore: "In the same place, the exiled Aeschylus, / forlorn, measured his verses and lines, in that parched gulf the eagle saw him, / and it was his last day" (45; Là Eschilo esule / misurò versi e passi sconsolati, / in quel golfo arso l'aquila lo vide, / e fu l'ultimo giorno). According to the legend, which tells a fascinating story of human-nonhuman interaction and confusion, he died because an eagle, in order to break the shell of a turtle it held in its claws, dropped it on Aeschylus's head, mistaking his skull for a rock. This literary reference justifies the "poetic fervor" and "historicist lyricism" (Banita 2014, 147–48) of the narrative accompanying the film's images and sound.[29]

In addition to the documentary's tragic tone, a funereal and anguished message ultimately emerges from the unspoken but still spectrally present remaining verses of Quasimodo's poem:

> Man of the North, / you who want me / reduced or dead for your peace, just you wish: / my father's mother will be one hundred years old / next spring. Hope that tomorrow / I shall not be playing with your skull, yellowed by the rains.

> Uomo del Nord, che mi / vuoi / minimo o morto per tua pace, spera: / la madre di mio padre avrà cent'anni / a nuova primavera. Spera: ch'io domani / non giochi col tuo cranio giallo per le piogge. (1966, 45)

In a historical moment and geographical context in which ENI, an oil company from Northern Italy—after "dropping" onto this southern region (like the "killing turtle" on Aeschylus's head)—has been settling in, drilling and building a petrochemical plant on those "straw-colored sands," the reference to a "Man of the North" who wants the poet "reduced, or dead for [his own] peace" acquires an additional layer of meaning. Far from eliciting feelings of appreciation or admiration for this Northerner, the poet ends his text by expressing a desire to outlast and even take some form of revenge on him ("Hope that tomorrow / I shall not be playing with your skull"). Thus despite its official propagandistic message, Sciascia and Ferrara's decision to open their documentary with this poetic reference immediately affects its overall tone and form, complicating any straightforward interpretation or simple designation of it as a typical corporate film.

Alternating sequences related to the digging and construction of the refinery with others depicting archeological finds, religious celebrations, folkloric rituals, and traditional craft-making, the film strives to maintain balance. On the one hand, we have the new, required celebration of petroleum as a nonhuman agent of cultural and material transformation, only visible via the technological marvel that is drastically reshaping the local land- and seascape. On the other, we find the alternative, socially conscious recollection, appreciation, and celebration of a human dimension that, as an integral part of the same environment, is also dramatically changing, if not at risk of completely disappearing.

Maintaining an open dialogue between past and future, between the desire to improve the local standard of living through industrial development and the resistance to losing ancient roots and traditions, the film's opening archaeological shots unveil a quintessential "landscape of [material and semiotic] porosity" (Iovino 2014, 4). This is a place where not only do dreams connect the world of the dead with the living, revealing to the latter where to find "the sites of buried wealth" (the so-called *trovatura*), but, as we are told, the material remnants of the ancient Greek settlement discovered in the excavation sites ("the columns, the coins, the urns, the jewels, the layouts of ancient Gela") surface directly "in the human heart" (Sciascia and Ferrara 1964) and, acting together with it, generate new narratives and interpretations.

In this particular Mediterranean scenario of hybridity, as Elena Past observes, "energy and culture . . . intermingle . . . petroleum moves through vertical layers of space and, in a parallel movement, geological layers of time" (2016, 377, 380). In one of its initial sequences, the film tells the story of an elderly local who, inspired by a dream, dug out a section of the ancient walls (which, he tellingly states, are now "a part of his life") and then was able to "re-create in his mind old war facts" (Sciascia and Ferrara 1964), ultimately weaving together and performing an oral tale about the Carthaginian and the American landings in Sicily.[30] This sequence stands out as a clear example of the connection linking places, bodies, matter, imagination, and meaning. Or, as Serenella Iovino states, it perfectly exemplifies "the way the world enters and conditions habits of living, thus determining the way living beings *in-habit* the world . . . the enactive processes of 'embodied knowledge' . . . rooted in the mutual porosity of bodies and world" (2016a, 101–2).[31] So if the ancient walls and archeological findings inspired and meaningfully shaped the old man's story and life, what kind of hybrid narrative can we expect petroleum to tell?

The encomiastic song to science sung by a local songwriter at the end of the documentary shows that oil too, as a form of modern underground treasure, can narrate its own porous stories. But if the narrative generated by the archaeological findings could surface from a "sand/dust that is still a measure of human time," the one produced by this new, much larger wealth emerges out of a much deeper geological time, thanks in part to precise machines that take the place of the revealing power of human dreams. Thus if the distance between human and petroleum histories is shortened by its presence and organic origin under different layers of sand, then we are also aware that this new chapter of oil drills and machinery will be an essentially dehumanizing one. In this sense, the image of a lonely palm tree that appears amid the construction area—"saved as a symbol of an ancient Mediterranean lifestyle" (Sciascia and Ferrara 1964) and yet progressively isolated and surrounded by mountains of iron, steel, and concrete—eerily anticipates and embodies the seclusion, sense of displacement, and vulnerability that humans, together with the local flora and fauna, will soon be facing in this newly engineered landscape.

However, viewers are told that human beings still represent a crucial part of this picture of progress. While mostly aerial, panoramic shots focus on the new plant's equipment and components, only incidentally including the workers, the voice-over narrator observes, "Human proportions seem reduced; the man laboring becomes a small thing." The more-than-human (and inhumane) agency of all these autonomous objects and pieces of equipment becomes increasingly apparent as one looks in detail at the enormous coke combustion chamber that "slowly, with meticulous and careful work, *inserted itself* into the landscape and, from now on, will be a part of it, like the dunes and the trees before" (emphasis added) and at the "machines that stick into the sand the poles for the wharf" (Sciascia and Ferrara 1964) or set the tank farm in place.[32]

Struggling to find a more tangible (and visible) human dimension and, together with it, "a measure of oneself," the camera then moves outside of the refinery and sets its focus on the town's festivals, rituals, and religious traditions as well as on the work of artisans and manual laborers who are filmed while building Sicilian carts or manufacturing tiles in an old brick factory. Nonetheless, toward the end of the documentary, in a series of incisive intercuts and rapidly juxtaposed shots, Ferrara cannot help but emblematically underline the extent to which human bodies are entangled with their oily surroundings and the inextricable, almost neurotic continuity and permeability between inside and outside, town and plant, the human dimension and petroleum itself. He conveys this impression first by showing a local but also universal workman, "no different in appearance and behavior from a Piedmontese or Lombard worker" (Sciascia and Ferrara 1964), leaving his job at the end of the day. Ferrara then films this worker again as he walks in a religious procession in town holding a (presumably oil-lit) torch in his hand. This impression is amplified by intercutting footage of drummers performing music in the town band with footage of workers hammering (that is to say, making the same banging gesture) on some enormous oil pipes. Lastly, he associates the images of a candle and the lights of the festival at dusk with those of the burning flame on top of a combustion tower and the illuminated plant at night. The message is clear: in this maritime Sicilian place of "myth, matter and liquid flows" (Past 2016, 375), not only are borders and identities porous, and past traditions and gestures intermingled with present ones, but human beings and petroleum are indissolubly linked, just like industry and culture are ultimately fueled by the same substance.[33]

Far from being "imaginatively sterile" (Ghosh [1992] 2005, 138–39) or absent from fiction, petroleum is here intimately involved with (if not actually shaping) the same life stories, rituals, and belief system of the population. Or, as Karen Pinkus states (paraphrasing Reza Negarestani's *Cyclonopedia*) in her book *Fuel*, "Oil . . . is not only the lubricant of narrative, it is narrative, the only narrative, all narratives" (2016, 76).[34] After having previously sung "ancient stories about pride, passion, and honor" (Sciascia and Ferrara 1964), the conclusive song to science—now with its latest references to the enlightening role of science and machines, to petroleum as a nonhuman agent able to change lives for the better, to the land, to the settling moon, and to the rising sun—captures the "stratified natural-cultural ecologies of this place" (Iovino 2014, 103). An apparent ode to optimism, the song points toward a path that, thanks to hydrocarbons, will allegedly go from darkness toward light, from resignation toward hope, from poverty toward prosperity.

However, if the words of the song tell a story of a bright future, the documentary's final image of a burning flame atop an oil tower cannot truly erase the sense of disorienting material and metaphorical darkness that surrounds the plant and the village that hosts it. The absence of light and words in the film's final sequence, especially when compared to its sun-drenched, poetry-filled opening, creates a conceptual and cognitive disjunction. This last frame seems to suggest a dynamic of estrangement—one that Iovino, quoting anthropologist Ernesto De Martino (1908–1965), describes as a "territorial anxiety" pointing to the loss of "a storied landscape, a densely inhabited world representing 'the living history of the other in us'" (2016a, 100). At the same time, as this sequence also points to the local actors' inability to see and read what is going on in their territory (i.e., to a loss of legibility of the landscape), it also reminds us of the necessity, theorized at length by geographer Eugenio Turri in his *Il paesaggio come teatro* (Landscape as theater), of "reclaim[ing] the role of spectators after decades of transforming activities" (1998, 18; reclamare il ruolo di spettatori dopo decenni di attività transformative) in order to prevent and resist the destruction of our landscapes. It hints, in other words, at how it is increasingly difficult, when one can only "act inside" a landscape and is unable to get "outside" of it,[35] to be real "spectators able to judge and obtain useful teachings from the spectacle" (130; spettatori capaci di giudicare e di ricavare utili insegnamenti dallo spettacolo).

By providing us with a much-needed outside perspective and reminding us of our own roles as spectators, Ferrara and Sciascia's documentary acquires an important enlightening role that tensely accompanies its propaganda function and, in helping us see beyond the latter, avoids being just an instrument aimed at facilitating forms of corporate blinding. Sciascia himself, reiterating the cognitive, cultural-ecological function of so much of his work, effectively summarizes a similar dynamic a few years later in his short story "The Wine-Dark Sea" (1985), in a fictional conversation between Mr. Bianchi, an engineer from the northern town of Vicenza who is traveling by train to Gela in order to start working in the plant, and the indigenous, homebound Professor Miccichè:

> "You have your oil now," Bianchi said by way of consolation. "Oil? Believe me, they'll soon grab it," said Miccichè. . . . "They'll grab it . . . one long pipe-line from Gela to Milan and they can just drain it off." (Sciascia 1985, 45)

> "Ora avete il petrolio," disse l'ingegnere, a consolarlo. "Il petrolio? Mi creda, se lo succhiano," disse il professore. . . . "Se lo succhiano . . . una canna lunga da Milano a Gela, e se lo succhiano."

Professor Miccichè's compelling description of a vampiric Italian North that drains the resources of the South without contributing to the South's long-term socioeconomic and cultural development is a familiar story that, while it resonates at the global level, may also be traced to Italy's postunification years. This was a period when crucial decisions regarding the overall industrial growth of the country were made, and its whole Mezzogiorno and islands—far away from the House of Savoy and the pulsating Piedmontese heart of the new nation—began to be viewed as a peripheral territory that was to serve the needs of rising liberal capitalism.[36] Alluding to this complex and controversial history helps underline how Miccichè's words illustrate similar exploitative dynamics in other parts of Italy. The region of Sardinia, acquired by the Duke of Savoy in 1720 and a Savoyard possession until after the Italian unification (1861), happens to be one of these places.

In the final section of this chapter, let us then keep in mind Sciascia's image of the sucking pipeline in the mouth of the North but substitute Gela in Sicily with Sarroch (San Rocco), near Cagliari in the southwestern part of Sardinia, as a parallel example of an island-based, stratified natural-cultural landscape that, from the early sixties on, has been radically transformed by the presence of what is now the second-largest petrochemical plant in Europe. Following a now familiar pattern, the Milan-based Saras (Società Anonima Raffinerie Sarde), owned by the Moratti family, purchased the land—originally 450 acres along a fertile, coastal area traditionally dedicated to agriculture, fishing, and sheep farming—and began building the plant in 1962, invoking a progressive rhetoric of necessary economic growth and development for both the territory and its inhabitants. Today the Saras/Sarlux industrial complex extends over 2,000 acres, refines more than three hundred thousand barrels of oil per day, and uses the corporate motto "Safety Is Our Energy." This motto is, at the very least, contradictory if only because, according to the European Pollutant Release and Transfer Register (E-PRTR), in 2017, Sarlux released over 6 million tons of carbon dioxide, 3,250 tons of nitrogen oxide, 4,310 tons of sulfur dioxide, 60.6 kilos of arsenic, 215 kilos of copper and compounds, 1.1 tons of nickel and compounds, 14.1 tons of benzene, and 155 tons of PM10 (particulate matter; E-PRTR 2017).

As it turns out, what actually grew and developed over the years—even more than the size of some bank accounts, the consumption of goods, or the international flow of crude from Libya, Russia, the North Sea, and the Middle East—was the release of pollutants and contaminants in the once pristine land and sea surrounding Sarroch and an anomalous number of tumors and other diseases among workers, animals, and local residents. This is a dynamic that, to quote Heather Houser in her *Ecosickness in Contemporary U.S. Fiction*, underlines once again the "homologies between environmental and somatic vulnerabilities" and the "shared endangerment . . . of contemporary bodies and environmental systems" (2014, 3–4). While I wish to expand Houser's neologism beyond the field of fiction (as I include a documentary film), here I closely follow and share her definition of sickness as "a pervasive dysfunction [that] cannot be confined to a single system and links up the biomedical, environmental, social and ethicopolitical; and it shows the imbrication of human and environment." At the same time, and even though I share the idea that the emotions caused by literature (as well as film and other artistic forms) contribute to our ethical and political stances, I do not exclusively focus on the affective mechanisms in narratives that "bring readers to environmental consciousness" (11).

What eventually emerged out of this complex Mediterranean landscape, together with the hybrid stream of oil and health problems, were a number of creative works from different artistic media that represent the commingling between "people and things . . . matter and meaning" (Past 2016, 370) and the hazardous human-environment interaction occurring in this part of Sardinia. These cultural artifacts offer a counterdiscourse about the alleged benefits resulting from the industrial modernization of the island.[37]

In the final pages of this chapter, I engage with a documentary film and a novel, both set in and focused on Sarroch, that address different yet relatable forms of "ecosickness" in this area and bring to light oil's sociopolitical and ecological costs and, with them, "the struggle to know the invisible risks that travel across bodies and landscapes" (Alaimo 2010, 23). The first is Massimiliano Mazzotta's award-winning independent documentary film *Oil: La forza devastante del petrolio; La dignità del popolo sardo* (Oil: The devastating force of petroleum; The dignity of Sardinians; 2009).[38] The second is Francesco Masala's novel *Il parroco di Arasolè* (The parson of Arasolè; 2001).

Similar to other recent international documentaries that feature the problems surrounding oil, Mazzotta's film may also be considered as a form of ecocinema and "political pedagogy that not only shape[s] audience understanding of the issues in question but also hope[s] to generate political and ecological responses that otherwise would not occur" (Szeman 2014, 350–51).[39] If, as Christopher Schliephake observes, drawing from the work of German philosopher Siegfried Kracauer, films in general "are primarily made for the portrayal and disclosure of physical reality," Mazzotta's documentary in particular fits the description of those that are able to reveal "the interactions and correlations between bodies, materials, and fluids that exist in a specific environment" (2015, 144–45). As it tells the story of the Saras oil refinery plant in Sarroch, the film also underlines the internal contradictions and stubborn persistence of an industrial development model that, while creating some jobs, has largely ignored its damaging long-term effects and "slow violence" (Nixon 2011) on both the local population and territory.

Despite its explicit title, *Oil* reflects a common feature and limitation of the documentary medium—that is to say, the aesthetic difficulty of capturing "a socially ubiquitous substance that remains hidden from view" (Szeman 2014,

364).[40] Thus while throughout the film we repeatedly see panoramic images of the enormous industrial conglomerate at Sarroch—with its skyline made of pipes, tanks, containers, and smokestacks emitting smoke and fumes of different colors and density—oil itself, as the material substance responsible for this complex industrial apparatus, is never in sight. But as one soon learns, Mazzotta's real interest is less in capturing oil on camera, addressing the global issues of the energy crisis, or treating oil as "something stranger, full of metaphysical mystery and subtlety" (Szeman 2010, 43) than in exposing, more in narrative rather than in visual terms, its multiple toxic agencies—its very real and concrete effects on bodies and the local (human and more-than-human) landscape—and eliciting questions of environmental justice and risk management.

The film has three main objectives and narratives. First, it evenhandedly documents the health problems of the mostly blue-collar workers at the plant by directly interviewing them and presenting their testimonials together with supporting and denying claims of other witnesses, including local politicians, doctors, and plant managers. Second, it aims to clear up any doubts about the direct link (or literally, "crude relationship") between Saras's presence in this territory and various forms of disease, repeatedly suggesting that the Moratti family, who originally built the Sarroch plant, is largely responsible for the devastating situation depicted. And third, taking Saras/Sarlux as an example, it shows some of the larger, "systemic" (Szeman 2014, 356) faults, limitations, and impasses of criminally blind administrative and economic policies at both the regional and national levels that contribute to the current "sicknesses" affecting the community and the environment within and beyond Sarroch.

Although the documentary mostly focuses on the victims of the oil refinery—on the older people who have already suffered health problems and the younger workers who are aware that they are endangering their health by working there—its concerns also move beyond the specific human dimension and ultimately address the toxicity and vulnerability of a much larger, traumatized ecology. After an initial sequence in which original newsreel footage combined with voice-over commentary retraces the inauguration of Saras in 1966 under the auspices of Giulio Andreotti (at that time, the Italian industry minister), "Ambiente" (Environment), the first among the ten chapters of the film, immediately brings up the dramatic "trans-corporeal processes" (Alaimo 2010, 104) that are active in this area. Here we learn from interviews with various locals about the widespread flow of toxic substances from the oil refinery to the air, land, and sea of the Golfo degli Angeli—which is said to have "the most fish-teeming waters in Sardinia"—and into marine life, food, and human and animal bodies. Two of these interviews are particularly revealing. The first is with a local spear fisherman who begins by tentatively saying, "Yes, maybe there's a little something in the water." He then remembers that a few days earlier, he caught "a diesel-oil-flavored sea bass" (Mazzotta 2009; spigola-diesel) fifteen kilometers away from the refinery. The second interview features a young shepherd who talks about the amount of pollution in the ground and in the plants around the industrial site and about his recent need to dispose of a sulfur-smelling lamb he had slaughtered because its meat was inedible.

In sharp contrast, these disturbing accounts are followed by a cut to a 1959 archival clip displaying a grinning Angelo Moratti telling an audience of investors and local politicians an anecdote about one of his engineers who lost a shoe in the mud while surveying the land for construction. Moratti ends his story by recalling how a member of his team interpreted this event as a lucky omen, since it meant that Saras had already been "incorporated into the Sarroch environment" (Mazzotta 2009). His remark, however, becomes increasingly alarming as *Oil* progresses, and viewers learn more about the actual ecological footprint left by Saras (in the mud and beyond) and, in turn, the hazards presented by highly toxic filter cakes, lethal gases that become odorless when highly concentrated, poisonous dust, and the related tumors, respiratory illnesses, DNA-modifying diseases, and deaths affecting every form of life there.

Imre Szeman identifies three main narratives that typically articulate the "problem" of oil and its future: "strategic realism, techno-utopianism and eco-apocalypse" (2014, 358). While there are few doubts that the latter label (eco-apocalypse) is the one that most closely describes the situation presented in Mazzotta's *Oil*, the director's balanced approach, as he alternates emotion-filled scenes with factual scientific and medical information, also prevents rigid categorizations. His reflexive ethical stance and activist commitment show the extent to which the area's landscape and inhabitants—"human and nonhuman, marine and terrestrial" (Past 2016, 379)—are affected by the petroleum industry. So while the last, powerful, and undoubtedly apocalyptic image of the documentary is that of a shepherd wearing a gas mask while his sheep graze in a field next to the refinery, the data-rich captions that follow also inform viewers that some positive changes to limit the amounts of emissions either are underway or will soon be implemented at Saras. This sort of documentary is, like all cinema, "intricately woven into industrial culture and the energy economy that sustains it" (Bozak 2012, 1). Yet it is also able to "interface with and shape our imaginations of the material environment" (Rust, Monani, and Cubitt 2013, 16) and thus is likely to have played a role in pushing the local government to pass new regulations about emissions. This alone gives reason to avoid fruitless nihilistic positions.

Saras sued Mazzotta, claiming that the film damaged its corporate image, but the legal battle was eventually resolved in favor of the director, since it was recognized that his work is of national public interest. Thus not only do we see how Mazzotta's work "counters petroleum energy with Mediterranean cultural energy" (Past 2016, 381), but we also see, along with Hubert Zapf, the potential that this kind of awareness-raising work, in parallel with the sick bodies it exposes, has to function as "a sensorium for what goes wrong in a society" and "for the survival of the cultural ecosystem in its long-term co-evolution with natural ecosystems," and vice versa (2016, 91, 26).

As previously mentioned, Sarroch—or "Sarrok," as Don Adamo, the psychologically tormented, existentially confused, and "polluted" (and eventually, "polluting") Catholic priest and narrator in *Il parroco di Arasolè* calls the new industrial town where he has been transferred from the rural village of Arasolè— is also at the center of this compelling novel by Sardinian writer Francesco Masala (1916–2007).[41] Set in 1961 during Italy's "miracle years," this diaristic, parable-like soliloquy mixes historical facts and fictional autobiography and is neurotically narrated in the first, second, and third person. Complementing Mazzotta's documentary, it reflects, at the levels of both form and content, some of the disruptive environmental changes, sociocultural degenerations, and as Pasolini teaches us, anthropological mutations that accompanied Italy's— specifically, this southeastern part of Sardinia's—process of modernization and conflictual negotiation with consumerist, neocapitalist models.[42] At the origin of such mutations and degenerations, which may again be perceived as concrete pathologies and diseases affecting both places and people (the latter at the physiological and spiritual levels), is petroleum itself—this new "god coming out of the depths of hell, the oil coming out of stone" (Masala 2001, 55; un dio uscito dal profondo dell'inferno, l'olio di pietra). Similar to other elements of matter that possess narrative agency and interact with the bodies of living organisms and their physical settings, this substance replaces traditional beliefs and cultural values, simultaneously revealing and obfuscating meanings, questioning identities, and reshaping territories, narratives, and ontologies. Oil's pervasiveness and its hybrid mix of nonhuman vitality and animal fertility surface in one of the narrator's frequent "cerebral onanisms" (45; onanismi cerebrali):

> Crude oil comes out of the tankers' belly, black and yellow like a viper's eye; it flows cold within the pipes, then it warms up its veins in the distillation ovens, achieves orgasm in the coils, couples inside the distillation towers like a filthy beast with thousands of sexes and, finally, it gives birth to scores of children: gasoline, vaseline, glycerine, paraffin, methane, butane, hexane, octane, ethylene, acethylene, propylene, polystyrene, alchylates, nitrates, chlorinates, sulfonates, et cetera, et cetera.
>
> Il petrolio grezzo esce dal ventre delle navi petroliere, nero e giallo come l'occhio della vipera, scorre freddo dentro i tubi, va a scaldarsi le vene nei forni di distillazione, entra in orgasm nei talami a serpentina, si accoppia come una bestia immonda dai mille sessi dentro le torri di frazionamento e, infine, partorisce migliaia di figli: benzina, vaselina, gliderina, paraffina, metano, butano, esano, ottano, etilene, acetilene, propilene, polistirene, alchilati, nitrati, clorati, solfonati, eccetera, eccetera. (17–18)

As this vivid description of the postextraction half of oil's life cycle demonstrates, Masala's narrative blurs the boundaries between body and environment, the organic and the inorganic, the animal and the material via the trope of oil, offering insights into the inextricable links among land, substances, human and nonhuman beings, and petroleum and its derivatives' ultimate responsibility as agents of decline and of natural-cultural degradation.

The novel opens with the image and description of a total solar eclipse—a "manifestation of nature" (Masala 2001, 7; manifestazione della natura) and nonhuman event that simultaneously refers to the parallel, threatening obscuration of the natural and cultural landscape in Sarroch and to the "mental uneasiness [and] interior malaise" (7; disagio mentale [e] malessere interiore) that is affecting the protagonist:

> I am forced to take into consideration . . . that there is, also, some kind of relationship between the darkness that is slowly covering the earth and the shadows of my soul. . . . From the top of the bell tower, Don Adamo looks at the obscene conjunction. The sun and the moon are lining up perfectly, like a single, monstrous hermaphrodite. The bulge of the chromosphere looks like the red swelling of a cosmic venereal disease.

> Sono costretto a prendere in considerazione . . . che ci sia, anche, qualche relazione tra l'oscurità che va lentamente coprendo la terra e le ombre della mia anima. . . . Dall'alto del campanile, Don Adamo scruta l'oscena congiunzione. Il sole e la luna perfettamente combaciano, come un unico, mostruoso ermafrodito. La protuberanza della cromosfera sembra il rosso bubbone di una cosmica malattia venerea. (8, 13)

The eclipse Don Adamo beholds from Sarroch's new bell tower represents a natural phenomenon, a cosmic embrace, and the subsequent shadow falling over human and celestial bodies. As it emblematizes a literal and metaphorical spreading of darkness over the environment, it also refers to the progressive concealment and erasure of the local Sardinian dialect due to the imposition of standard Italian. As Masala puts it, "The language of the defeated, of wheat, grass and sheep" (2001, 92; La lingua dei vinti, la lingua del grano, dell'erba e della pecora) is overcome by "the language of winners, of oil and tar" (92; la lingua dei vincitori, la lingua del petrolio e del catrame). Similar to an ever-expanding oil slick, the eclipse destabilizes univocal meanings and questions rigid ontological distinctions between the dimensions of the spiritual and the material, the natural and the cultural. In short, it points once again to the entanglements and permeability between bodies and world, as encountered in Mazzotta's documentary and in many of the texts I consider in these pages.

Most significantly, this astronomical event evokes the imbrications that exist between outside and inside, between cosmic, nonhuman events and human corporeality—not to mention the connection of these dimensions to those textual areas that, in strict correlation with the literal and metaphorical poisoning that is happening in the landscape, are similarly obscured and "poisoned" by the narrator's "tarry" language and multilingual confession made of "*inventio ambigua . . . dispositio incomposita* [and] *elocutio spuria*" (Masala 2001, 87).[43]

In defining *pollution*, Iovino reminds us that "pollution and poisoning are mental as well as physical phenomena, subjective as well as objective phenomena" and that "pollution is an interplay of harmful material substances and harmful discourses and practices" (quoted in Adamson, Gleason, and Pellow 2016, 168). I would add that "discourses and practices," in turn, may also become corrupted precisely because of those "harmful material substances." Don Adamo's increasingly dirty language, expressive unreliability, and impure and inappropriate behaviors—culminating in a corporeal, physiological form of pollution following his move to an oil-drenched Sarroch—would seem to confirm this suggestion. Simply put, oil, together with the sort of "harmful discourses and practices" that are so effectively illustrated in Mazzotta's documentary, contributes to pollution in Sarroch, and then such pollution has the power to taint previously harmless or relatively wholesome practices and discourses, such as those of a Catholic priest in Masala's novel. This vicious circle further corroborates the mutuality between the dimensions of the discursive and the material—the "dance of matter and meaning" (Iovino and Oppermann 2014, 9) in this place. One should not forget, however, that if discourses can be negatively affected by some poisonous agent or toxic colonizer, they can also shape narratives that may function as potential antidotes and enlightening alternatives to such a polluted state. As Cree scholar, poet, and visual artist Neal McLeod puts it, "Storytelling . . . becomes a lens through which we can envision our way out of cognitive imperialism, where we can create models and mirrors where none existed, and where we can experience the spaces of freedom and justice. Storytelling becomes a space where we can escape the gaze and the cage of the Empire, even if it is just for a few minutes" (quoted in Luisetti, Pickles, and Kaiser 2016, xiv).

That said, if in this novel plurilingualism and various lexical discontinuities become the only ways for the narrator to "express the multi-dimensionality of the real" (Maniowska 2012, 321), then linguistic communication is here far from being straightforward, fruitful, complete, truthful, or even really possible. Since the protagonist priest is supposed to be the spokesperson of a metaphysical and transcendental truth, this situation could not be more ironic. In the following lines, Don Adamo succinctly describes his relationship with some parishioners who were forced to migrate to Germany after having been "expelled by Sarrok's cybernetic refinery" (Masala 2001, 67; espulsi dalla raffineria cibernetica di Sarrok) and failed by its promises of comprehensive socioeconomic improvements: "Here, this is how it is: we all speak the same language and yet we still cannot understand each other" (70; Ecco com'è, parliamo la stessa lingua ma non ci comprendiamo lo stesso fra di noi).[44] Significantly, the text abounds with metaliterary references to various forms of perversions and improprieties that, once in Sarroch, affect the protagonist's mental and spiritual health and his expressive capacities:

> It often happens to me, too often, that I like to hear myself talk . . . it is a form of cerebral onanism, a sort of linguistic perversion, a way to communicate that is similar, very similar, to the vice of making love to yourself.

> Mi capita, spesso, troppo spesso, di parlarmi addosso ... una forma di onanismo cerebrale, una specie di perversione linguistica, un modo di comunicare simile, molto simile, al vizio di far l'amore con se stesso. (9)

> His monstrous metaphors, like Sarrok's oil, pollute everything they touch.
>
> I mostri delle sue metafore, come il petrolio di Sarrok, inquinano tutto quello che toccano. (27)

> I could no longer talk. I had become voiceless.
>
> Non ce la facevo più a parlare. Ero diventato afono. (77)

> His *ars dicendi* is like him ... [made of] words to signify everything and the contrary of everything.
>
> È simile a lui la sua *ars dicendi* ... parole per significare tutto e il contrario di tutto. (87)

> Actually, I am a big liar.
>
> In realtà sono un grande bugiardo. (98)

As hinted previously, sexuality plays a role in the protagonist's "onanistic" ruminations throughout the novel. This happens in particular when the narrator introduces Eva, the woman about whom he has been fantasizing since he was a child in Arasolè, and then a repulsively fat, drug-addicted prostitute—an allegorical image of a polluted earth and other poisoned corporalities or, arguably, even an anthropomorphic representation of oil itself—whom he tries, unsuccessfully, to carnally possess. Significantly, we learn that the prostitute is living "like a tree ... like a stone" (come un albero ... come una pietra) and contains within herself "an incomprehensible, dark thing" (Masala 2001, 77; una cosa oscura, incomprensibile), a remark that once again draws revealing parallels between female bodies and an elemental, nonhuman nature.

Reflecting the complementary poisoning of the exterior environment and, as Don Adamo puts it, of a "disrupted, polluted, amoral society" where "there is no earthly goal that rewards my chastity" (Masala 2001, 61; una società disgregata, inquinata, amorale non c'è alcun fine terreno che compensi la mia castità), sexuality too—along with languages, meanings, and knowledges—is progressively tainted by the oil plant.[45] In this ever dirtier context, the novel unsurprisingly reaches its provisional conclusion in conjunction with

> three orgasms: the asthmatic and filthy one of the poor priest, the desperate and incomplete one of his unlucky spouse [Eva], and the apocalyptic and happy one of the sun and the moon who terminated their astronomical embrace.
>
> tre orgasmi: l'orgasmo asmatico e turpe del povero prete, l'orgasmo disperato e monco della sua malfatata sposa e l'orgasmo apocalittico e felice del sole e della luna che hanno concluso il loro astronomico amplesso. (109)[46]

In the final passage, the "large stain of tar" (111; grande macchia di catrame) that Don Adamo sees on the ground at the bottom of the bell tower can be interpreted as not only the materially toxic pollution and symbolically murky outcome of those human and nonhuman embraces but also the objective correlative of the many dark spots, slippery allusions, and obfuscations of meaning present in the narrator's confession.[47]

By now, it should be clear that the main reason I have lingered over this connection between communicative and sexual issues is to emphasize how the natural, petroleum-like darkness brought by the eclipse corresponds to the cultural obscurity and linguistic and epistemological insufficiencies of the text and, in turn, to Don Adamo's spiritual, moral, and corporeal *descensio ad inferos*. Always more an "industrial priest . . . a plastic minister, a tarry parson" (Masala 2001, 61–66; prete industriale . . . un sacerdote di plastica, un prete al catrame) whose own sex has been "poisoned by the petrochemical industry" (80; inquinato dall'industria petrolchimica), the priest's body is thus enmeshed in the ever more contaminated, unavoidably entangled dimensions of nature, culture, oil, and narrative. At the very core of the narrator's communicative failure, imperfect knowledge, semiotic confusion, and frustrated sexual fantasies, there are oil and the oil refinery, material entities that, by literally eclipsing places and bodies as well as obscuring meaning, knowledge, language, and emotion, also represent "a breach, an epistemic rupture in the mind of this place . . . on the cognitive level" (Iovino 2014, 103).

As the exterior landscape in Sarroch becomes progressively poisoned, darker, and unintelligible, behaviors, language, and meanings meet a similar fate of sterility and emptying. Similarly, as the premodern world the protagonist and his rural parishioners once knew (Arasolè) is vanishing together with its communal traditions, religion, language, myths, and agropastoral sexual fertility, not only is their general well-being endangered and their expressive and cognitive skills reconfigured, but they also undergo a process of dehumanization that transforms them into alienated, objectified, posthuman tools: "One could become an ox in Arasolè but will certainly become a screwdriver in Sarrok" (Masala 2001, 20; Uno poteva diventare bue ad Arasolè ma divernterà sicuramente un cacciavite, a Sarrok).[48]

Seen in this light, Mazzotta's documentary and Masala's novel not only help illustrate the inextricable connections between the body—in both its material and spiritual components—and the place in which it is imbricated but also epitomize some of Karen Barad's ideas about the "intra-active" constitution of environments, bodies, and cognitive/discursive practices. Remarks such as "Matter's dynamism is generative not merely in the sense of bringing new things in the world but in the sense of bringing forth new worlds, of engaging in an ongoing reconfiguring of the world" (2007, 170) or "We know because we are of the world" (185) clearly reflect the situation in the novel if one simply thinks of oil's agency in erasing Arasolè and bringing forth a new industrial world. Furthermore, the realization that the pollution in Sarroch darkens some "practices of knowing and being" (Barad 2007, 185) while also revealing another "perturbing facet of [the protagonist's] existential prism" (Masala 2001, 62; faccia conturbante del mio prisma esistenziale) resonates well with petroleum's epiphanic potential to reconfigure the world at both the exterior and interior levels.[49]

Considered together, *Oil* and *Il parroco di Arasolè* thus deliver ecological narratives that highlight the imbrication among matter (in this instance, fossil fuel), human beings, and territory. As they both avoid "hiding" oil's problematic representation behind the "charismatic term 'energy,'" (LeMenager 2014, 185), they provide poignant accounts of the material-discursive disruptions and environmental trauma brought on by the neocolonial introduction of the Saras oil refinery in Sarroch and, by extension, in similar industrial contexts. By exemplifying two "critical and protestatory responses to oil" (Barrett and Worden 2014, xxvii), they signal the end of a chapter that has attempted to explore, both materially and aesthetically, some aspects and moments of Italy's slippery encounter with petroleum from the early twentieth century until today.

The stories these texts recount and that oil helped them conceive ultimately remind us of the decolonizing role that these kinds of narratives may have, as they urge us to reengage, rekindle, and resurrect critical "knowledges and sensibilities" (Luisetti, Pickles, and Kaiser 2016, xv) that many of us willingly forget or disavow in our daily routines of consumption.

Apua Ma(t)ter
Narratives of Marble

Vittorio . . . has been a stone breaker since he was nine years old. . . . The main part of his job is to listen to what the stones he finds have to tell him. He listens to them, he feels them by striking his hands over them, he cleans them, and he talks to them.

Il Vittorio . . . fa lo spaccapietre da quando aveva nove anni. . . . La parte principale del suo lavoro è quella di stare a sentire cosa hanno da dirgli le pietre che incontra. Le ascolta, le saggia passandoci sopra le mani, le ripulisce, e gli parla.

—Maurizio Maggiani, *Meccanica celeste*

Inside, marble is alive. It is like a plant that feeds off the mountain.

Il marmo, dentro, è vivo. È come una pianta, che ci ha alimenti dalla montagna.

—Marco Rovelli, *Il contro in testa: Gente di marmo e d'anarchia*

We are mixed, dissolved in a crucible / that is not ours, burned in an outburst / of marble.

Siamo mischiati, sciolti in un crogiolo / che non è nostro, arsi in uno slancio / del marmo.

—Ermanno Krumm, *Animali e uomini*

After having concluded the previous chapter by discussing oil's pervasive environmental impact and sociocultural effects in Sardinia, the present one takes us back to the Italian mainland. More precisely, we now go to the northwestern corner of Tuscany, at the border with Liguria, in the mountainous area of the Apuan Alps—famous for its marble production and the quarries that dot this territory—in order to explore another kind of material narrative.

In Jeffrey Cohen's short, meditative essay "Stories of Stone" (2010), his introduction to the volume of essays *Inhuman Nature* (2014), and his book *Stone: An Ecology of the Inhuman* (2015), he invites us to reconsider the supposed inertia and muteness of stone. Explicitly referring to the work of Bruno Latour, Stacy Alaimo, Jane Bennett, and Manuel DeLanda—all scholars who, from heterogeneous theoretical approaches, have been helping us rethink the notions of matter, agency, and the human—Cohen suggests that when viewed from a geological-temporal frame, stone is not just an immobile, nonsentient, and inanimate mineral. Rather, it is an active player in a variety of "geocoreographies." As such, he observes, "Stone moves. Stone desires. Stone creates: architectures, novelties, art.... Rocks ... like all life are forever flowing, forever filled with stories" (2010, 57, 63).[1] Stone, he adds, "enables movement and violence, extends cognition, and invites world-building" (2014, iii).

Cohen's remarks are validated right away if one thinks, for example, of the production of Italian artist Giuseppe Penone, a representative of the Arte Povera movement. To a larger extent than his other "poor" fellows, Penone's work has always been about individuating and conveying the metamorphic, animistic vitality of stone and the fluid relationship it entertains with the animal, vegetal, and human spheres. He is interested in representing humans' fundamental coextensiveness with nature as well as the confusion between entities and materials only apparently belonging to separate ontological realms. Art scholar Adele Tutter observes that some of the sculptures that constitute Penone's aptly named exhibition *Indistinct Boundaries* (2012) "look like ordinary tree trunks . . . covered with a thick coat of flat white paint save for certain areas where the bark is peeled away. . . . When one realizes that the painted bark is in fact sculpted from Carrara marble, and that the areas of exposed 'wood' are not wood but cast bronze, one is shocked in the way one is when one's perception and sense of reality are deeply challenged" (Tutter 2015). *Sigillo* perfectly exemplifies how Penone challenges traditional perspectives and, to quote Cohen, "extend[s] [one's] cognition" (2014, iii). It shows a horizontal veined white marble slab with a solid marble cylinder featuring veins in relief at its midpoint. This pattern underscores the unexpected analogy not only between marble and water, stirred and rippled by the wind, but also among the veins running through marble, those in a vegetable leaf, and those in a human body. In short, Penone's "organic sculptures" speak powerfully in favor of marble's vitality.

More recently, with her *Draped Marble* series (2015), contemporary conceptual artist Analia Saban echoes from the other side of the Atlantic Penone's reflections on material dynamism and his invitation to rethink ontological boundaries. By managing to drape a folded slab of marble over a sawhorse (Fiori di pesco; 2015), Saban invites us to both question the material's inert, rigid structure and contemplate its mutability and textural elasticity over time. As Kara Thompson observes in her insightful discussion of Saban's artwork, "When Saban turns a slab of marble into a fractured and folded blanket, the draped marble *is* and *represents*—in a Deleuzian sense, it is both being and becoming" (2019, 43).

Thus among the many varieties of stones and rocks that one could consider to further test Cohen's insights within an Italian-Mediterranean context, the Apuan (Carrara) marble ($CaCO_3$) and the geographical territory it comprises are undoubtedly fit for the task. Similar to many other stones, this quintessential Italian material is the product of the progressive, dynamic sedimentation and metamorphoses of both organic and inorganic substances that once were under a primordial sea—a process that, once again, challenges marble's static and inanimate nature.

Fabrizio Arrigoni reminds us that roughly 240 million years ago, the Apuan region was a "platform jutting out from the Euro-Asian continent, which was still fused with current continents to form the so-called 'Pangea,' covered by a single large primitive ocean" (2009, 20), while Marco Rovelli observes that

> marble is teeming with marine lives, carbonate sediments produced by past coral reefs, made of algae, sponges, corals. Those strata [are like] . . . under skin veins which the millenary work of the caves has discovered and brought to the surface . . . those marble strata are alive, deeply alive, and the whiteness that dazzles you is like a hypnotic concentrate of life.
>
> il marmo è un brulichio di vite marine, sedimenti carbonatici prodotti in quelle che furono scogliere coralline, dov'erano alghe, spugne, coralli . . . quegli strati . . . vene sottopelle che il lavoro millenario delle cave ha scoperto e portato in superficie . . . quegli strati di marmo sono vivi, profondamente vivi, e quel biancore che ti abbaglia è come un concentrato ipnotico di vita. (2012, 28)

Marble features prominently in and intertwines with human culture, history, art, and architecture. Its widespread utilization from Roman times through the Middle Ages and the Renaissance until today—when its extraction raises serious social and ecological concerns among environmentalists, intellectuals, artists, and regional administrators alike—suggests that it is "filled with stories" (Cohen 2014, iii).[2] For example, Alberto Asor Rosa, one of Italy's foremost public intellectuals, tells a simple yet relevant environmental tale. Pondering on the numerous risks associated with marble extraction practices, he remarks, "If the excavation policy that has established itself in the latest years continues, there is a disaster looming on the Apuan Alps. By extracting up to 4 million tons of marble per year, there is a risk of causing a historical, ecological and environmental catastrophe" (2014).

This call for wider public attention has not gone unheard. The two recent nonfiction books addressed in the following pages—Marco Rovelli's *Il contro in testa: Gente di marmo e d' anarchia* (Resistance in the head: Marble people and anarchists; 2012) and Giulio Milani's *La terra bianca: Marmo, chimica e altri disastri* (The white land: Marble, chemistry and other disasters; 2015)—discuss precisely how quarries have been contributing to the degradation of the local natural landscape, flora, and fauna (e.g., mountaintop removals, waste deposits, abandoned pits, and rivers and aquifers polluted by debris, fine marble dust, and hydrocarbons). They show how the Apuan quarries, together with other marble-generated industrial sites, have been affecting the health of local populations both physically (by exposing them to sound pollution, particulate, and chemicals) and financially (since only a small number of foreign-owned corporations reap most of the benefits of this "white gold"). Notably, these texts follow the path carved more than forty years ago by Pier Paolo Pasolini, who, together with Lotta Continua's Giovanni Bonfanti and Goffredo Fofi, denounced this problematic situation in his militant documentary *12 Dicembre* (12 December; 1972). In this unfinished short film, local marble workers are interviewed about the unhealthy conditions in the Apuan quarries and the many job-related deaths.

Consistent with the ecomaterialist framework informing this work, my objective in this chapter is to read these cultural stories and natural stories "diffractively"—that is, through one another—and "intra-actively" (Barad 2007, 33) in their unavoidable interrelations. As the cultural stories generated by human creativity and the natural stories produced by the nonhuman agency of materials coemerge, they interpellate and influence one another.[3] According to this perspective, which sees matter telling its stories through the imaginative efforts of human interpreters, the mineral, material landscape of Apuan marble becomes one whose physical, nonhuman narrativity, agency, and meaning can be explored and read in conjunction with the discursivity, agency, and meaning of human texts and expression.

To begin exploring how human imagination may give access to the accumulation of "storied matter" (Iovino 2014, 97) in this geographic area, I take a quick look at how the Apuan region has been (and still is) officially described and represented to the world through different media. Under the link "Massa Carrara" on the official website of Italy's Agenzia Nazionale Turismo (National Tourism Agency), for example, one reads, "The spectacular white marble caves in the heart of these mountains are unique in the world and reveal themselves as a gleaming vision before the visitor's eyes. Entering the mountain's heart, one can touch the same precious stony elements preferred by artists like Michelangelo and Canova." A subsequent click on the link to the "Apuan Alps" invites us to wonder in admiration and awe as we learn that "the quarries and the detritus look like sheets of snow" (Agenzia Nazionale Turismo, n.d.). Moving to the regional level, the rich and informative website of the city of Massa Carrara further emphasizes the "magnificent scenario created by this two-thousand-year-old human industry" and "the suggestive imposingness and peculiarity of these places" (City of Massa Carrara, n.d.). As the website lists the various kinds of quarries, it notes how some among these (i.e., the "gallery caves") are comparable to "immense cathedrals excavated in the heart of the mountains." It goes without saying that these inviting, attractive descriptions aim to promote this territory and to boost a business that has been a fundamental source of support for the local economy and the indigenous populations over many centuries.

In addition to these websites' references to illustrious artists of the past, cathedrals, and aesthetic beauty, they also manage to associate labor and leisure—that is, they attempt to transform excavation sites into tourist attractions. These sites, especially in the past decades, have become more industrialized, with advanced technology that has radically changed the nature of the marble-extraction business. In doing so, they perpetuate what environmental historian Stefania Barca, writing on the making of the Italian industrial landscape in the nineteenth century, has called a "pastoral ideal of industrialization, suggesting the total compatibility between nature and progress" (quoted in Armiero and Hall 2010, 217). In other words, in a way that resembles the dynamics Barca described, the marble quarries of the Apuan Alps have historically been widely perceived as products of "human industry" that—rather than constituting potential blemishes; ugly intrusions on a pristine, natural landscape; or agents of environmental disruption—appear to have enhanced the picturesque beauty and distinctiveness of the local nature through the years (at least until concerned groups began saying otherwise).

Among other disciplines and discourses, one then wonders if modern Italian literature, deliberately or not, has played a role in the construction of this industry-enhanced, aesthetically pleasing image and virtuous, idealized narrative of the Apuan Alps. A quick survey reveals that, paradoxically, some of the responsibility for this state of affairs could be assigned to a writer who has repeatedly displayed a profound environmental sensibility and attention to the disfigurations of the Italian landscape. Lombardy-born Guido Piovene, in his popular travel book *Viaggio in Italia* (Voyage in Italy), written in the years of the Italian economic boom, observes,

> Carrara rises between the beautiful coast and the dolomitic Apuan Alps. . . . An industrial zone was created before WWI. . . . Marble is still the main industry. I climbed up to the Apuan quarries, first by car, then in a small cableway. . . . It is a stupendous mountain, with a white, absolute light. Marble has a radiating power, a still splendor, even greater than ice. The snapdragons and the yellow daisies stand out on such whiteness. . . . And there is also a sense of joy in these quarries. Their white light causes in whomever visits them a kindling of one's thoughts, almost a sense of elation. The quarry where I stopped was wrapped in that reflection; some just-born chick, frightened by the mining, ran up and down in the courtyard.
>
> Carrara sorge tra la bella marina e le Alpi Apuane. . . . Una zona industriale fu creata nell'anteguerra. . . . La principale industria rimangono sempre i marmi. Sono salito alle cave delle Apuane, prima in macchina, poi nella piccola teleferica. . . . È una stupenda montagna: d'una luce Bianca, assoluta. Il marmo ha una forza d'irradiazione, uno splendore fermo, maggiori di quelli del ghiaccio. Risaltano, su quel bianco, le bocche di leone, le margherite gialle. . . . E c'è anche, in queste cave, un senso di gioia. La luce bianca delle cave porta, in chi vi passa, un'accensione dei pensieri, quasi un'esaltazione. La cava dove ho sostato era avvolta di quel riflesso; qualche pulcino appena nato e spaventato dalle mine, correva su e giú nel cortile. (1957, 329–30)

Even though, to be fair, Piovene also briefly mentions elsewhere in his account the dangers and the difficulty of the workers' jobs and observes the presence of anarchic tendencies in the area, his description of the quarry's splendid, icy magnificence as a place where technology, minerals, colorful flowers, and chicks joyfully coexist and the marble's whiteness enlightens and ignites the visitor's mind approaches the pastoral view of industry mentioned earlier.

That said, if we look at how the Apuan Alps territory and the operations related to marble extraction have been represented over the past century in the works of other Italian (specifically, local, indigenous Tuscan) writers and artists, one finds that they convey an alternative, much more complex and nuanced perspective of this area.[4] References to job-related injuries, death, displacement, exploitation, poverty, floods, and pollution, paired with an understanding of the social costs involved and of the interconnectedness of human and nonhuman matter, suggest that the early twentieth-century works by Ceccardo Roccatagliata Ceccardi and Enrico Pea and the modern texts of Maurizio Maggiani, Marco Rovelli, and Giulio Milani—despite their differences in genre, form, and objectives—share a deep, critical awareness of the impact and agency of the marble industry on both the territory and its inhabitants and of the material interrelations between them.[5] As a consequence, the texts discussed in the following sections problematize rather than contribute to, construct, or reinforce the official representations of the marble quarries, implicitly questioning any pastoral ideal of industrialization. They also provide a more realistic and environmentally conscious picture of the territory that they describe. Ultimately, all of them, in different guises and using different points of emphasis, question mainstream representations of these places and become sites of reflection and resistance to today's politics of marble extraction, processing, and marketing. Considering such politics as an additional form of what Heather Sullivan calls "dirty traffic" (2014, 97)—namely, one among the many deterritorialized, disruptive flows of substances, people, and bodies in the Anthropocene—and in light of her related reflections on the "dark pastoral" as a "trope of *exposure*" that helps reveal the hidden costs of industrial capitalism (2017, 26), let us then take a closer look at the definitively darker, antipastoral frame of the following texts.

"Walking on the Wild Side" with Ceccardo Roccatagliata Ceccardi

An appropriate point of departure to explore these mountains is to examine "Apua mater" ([1905] 1982), a sonnet written by the incredibly aptly named poet

Ceccardo Roccatagliata Ceccardi (1871–1919)—*Roccatagliata* literally translates as "cut rock"—dedicated to the places in the Apuane territory where he spent his youth (and that I shamelessly modified to title this chapter):⁶

O Madre terra, un popol di giganti	O Mother Earth, a people of giants
contro qual Nume libertà contese,	contended for freedom against which God,
ché il fier costume seppellìa co' franti	so that it buried its proud customs and its
penati in grembo del montan paese?	broken penates in the womb of the village?
Da le tue selve contro il ciel protese	From your woods leaning forward to the sky
e da' tuoi fiumi contro il pian croscianti,	and from your rivers roaring in the valley,
un'eco di racconti epici scese	an echo of epic tales descended
per gli evi, a' tetti de'nepoti e a'canti.	through the ages, to reach the homes and the chants of grandchildren.
Quindi grand'opre e libertà fermenti	You, Mother, are brewing great works and
tu, Madre, in cuore di impetuosa prole	freedom in the heart of your impetuous offspring who still feels despondent
che ne l'ombra dei Padri ancor s'abbatte:	in the shadow of the Fathers:
o nel pian le città morte co' lenti	either you gauge through slow oxen the dead
bovi misuri, o il grembo arduo di fratte	cities in the valley, or you weed a womb of steep bush,
ronchi, e palleggi i bianchi marmi al sole.	and rock the white marbles in the sun.

([1905] 1982)

First of all, in the footsteps of Ceccardi, I too plan to interrogate "Mother Earth" in order to listen to the tales that descended from its woods and rivers in the mountains to finally reach "the homes and chants of grandchildren." The personified "Earth" in this poem is expected to answer to the poet's initial question regarding a still vague, destructive struggle for freedom that took place in this historically conflict-rich, independent-thinking territory. Significantly, when Ceccardi writes that the rivers and woods are able to communicate with us and carry down from the mountains to the plains a number of their tales, he not only indicates the narrative natural-cultural wealth of this area but also translates into a concise, poetic form the ecosemiotical idea that landscapes are meaning-bearing, agentic texts in themselves and that they are "material narratives of a society's physical and cultural transformations" (Iovino and Oppermann 2014, 12).

Ceccardi romantically sees the maternal Earth of the Apuan Alps as a stirring source of creative inspiration and social and environmental information for its "impetuous offspring"—a source that continues to stir the rhythms of both human and nonhuman existence as it either "gauges" with the help of "slow oxen" some spookily empty "dead cities" in the valley or playfully "rocks the white marbles in the sun" (1982, 253). This is a place where a "womb" may simultaneously belong to the different corporalities of a mother, a "mountainous village," and some thick bushes (*fratte*). In the poem's representation of an extended nonhierarchical family constituted by human, mineral, vegetal, and animal characters (mothers, fathers, grandchildren, offspring, woods, bushes, rivers, valleys, cities, marbles, oxen), "Apua mater" thus underscores the particular porosity of this mountain landscape—namely, the mergings, meldings, and interplay of its semiotic, material, and sociohistorical components—and the necessity to interpret it in all its layered complexity. Our attempt to clarify the nature of the struggle alluded to in the poem—which, in turn, should also explain the reference to the "dead cities"—constitutes one more step in this direction.

In Ceccardi's earlier, mostly forgotten but remarkable text *Dai paesi dell'anarchia* (From the villages of anarchy; 1894)—a short, self-published pamphlet whose distribution was interrupted by the authorities because of its subversive and potentially explosive content—the poet passionately defends the cause of the Carrara marble quarrymen who, because of their vocal support for the democratic and socialist-inspired demands of the workers of the Sicilian Workers League (Fasci Siciliani) and their own acts of resistance and protest in Tuscany, were similarly targeted and violently repressed by the Crispi government.[7] Here Ceccardi argues against the imprisonment, condemnation, and subsequent displacement of young men who, like their Sicilian counterparts, were "tired of being exploited, of dying unknown under the rocks and the mines in the quarries for just a few cents a day" (1894, 6; stanchi di essere sfruttati, di morire per pochi centesimi al giorno—ignoti—sotto i massi e le mine delle cave), and whom "military law and a martial court had defined anarchist insurgents" (3; la legge militare e un tribunale di giberne avevan detto insorti anarchici). Most importantly, he gives us an accurate first-person description of the harsh life in Apuan marble quarries at the cusp of the nineteenth and twentieth centuries; of the socioeconomic, political, and class issues surrounding their operations; and of the extent to which the landscape of the caves commingles with the workers' bodies. The pamphlet is also an early example of (bio)regionalist and socioenvironmentally conscious writing:[8]

> I have been up there in Fantiscritti and in Ravaccione, the supreme quarries, in front of nature's immensity expressing itself in a strange form of rocky landscape, with its cyclopean shades, in front of the convulsive monstrosity of the mountains and the horridness of the *ravaneti*, [in front of] the bold peaks soaring in the sky, or losing myself in a white, bleary cloud with a blinding reverberation, and up there I said: men here work, they earn a living. While your foot sinks into the *ravaneto*, on your head hangs an always rolling boulder, and life is fragile when it is suspended from a rope attached to a simple picket. Oh, people from the bourgeoisie, whoever works here must be a titan and, let me say it, a hero, an old hero ... if one day, as it often happens, the rope will break and the rock descending from the quarries down to the square will change direction, if the white dust in a windy day will blind him, and a marble block blown by a mine will hit him, this man, if wounded, will have a ladder—four crossed pine posts—as his first bed and, if dead, his casket will just be a bag which used to carry gunpowders and mines, a fragile box carrying his smashed body.
>
> Io sono stato lassù a Fantiscritti e a Ravaccione, le supreme cave e dinanzi all'immensità della natura che si estrinseca in una strana forma di paesaggio roccioso, dalle tinte ciclopiche dinanzi alla mostruosità convulsa dei monti e all'orridezza dei ravaneti, all'audacia dei picchi svettanti nell'azzurro, o perdendosi in una bianca nube velata che acceca col suo riverbero, ho detto: gli uomini qui lavorano, ben si guadagnano il pane. Tanto il piede affonda nel ravaneto, tanto sulla testa è sospeso il masso che continuamente rotola, tanto la vita è fragile se attaccata ad una fune appesa ad un semplice piuolo che colui che qui lavora dev'essere un titano, ed almeno lasciatemelo dire, o borghesia, un eroe, sì, un vecchio eroe ... se un giorno, come spesso succede, la canapa della lizza si romperà, e il masso che scende dalle cave ai piazzali della marmifera, devii, se la polvere bianca di un giorno di vento lo acciechi e un blocco di marmo slanciato da una mina lo percuota, egli non avrà, se ferito, che primo letto una

scala, quattro pezzi di pino incrociati, e se morto, appena un sacco d'onde si asportò già polveri piriche, e mine, fragile cassa alle sfracellate membra. (8)

While rhetorically expressing through the then still popular aesthetic of the sublime his awe in front of the spectacle offered by the "immensity of nature," the "monstrosity" of these steep mountains, and the "horrid" accumulations of marble debris (*ravaneti*), Ceccardi also conveys the dangerous interpenetration of men and landscape in all of its physicality. With their feet sinking into the debris; their heads targeted by rolling boulders; their eyes blinded by the white, calcareous dust; their bodies hit by blown-up shards; and their lives dangling from a thin rope on a marble wall, men work here until their smashed corpses end up—reduced as just another discarded, no-longer-useful material—in the same bags used to carry explosives. And of course, while the author captures the violence that this fascinatingly beautiful and hostile territory throws toward men, he also exposes the even greater violence that men bring to this land:

> The rope often breaks and the enormous boulder—if men are not ready to escape, rolls onto them and takes its revenge, killing them: killing them, little guys who with their small means attempted to take him away from his holy native place.... Who has never seen a quarry, who has never dared climbing onto one cannot really have an idea. Just think that in the valleys of Canal Piccinino and Canal Bianco there are hundreds of them, one next to the other, one on top of the other. They look like enormous, candid wounds standing out of the widespread gray of these rusty-colored mountains. On the edges of steep cliffs, blood-like slots furrow over these quarries, wherever the rust hits the white.

> La canape spesse volte si spezza e il masso enorme—se gli uomini non son pronti a fuggire, rotola loro addosso e si vendica, uccidendoli: uccidendo essi piccoletti, che con piccoletti mezzi tentarono di portarlo via dal suo santo luogo natale.... Chi non ha veduto una cava, chi non ha osato salirci non può davvero farsene un'idea. E pensare che nelle vallate di Canal Piccinino e di Canal Bianco, esse si contano a centinaia, una dietro l'altra, una sovra l'altra. Sul diffuso grigio delle montagne arrugginite esse paiono enormi ferite candide. Cigli di rupi irte, scanellature di righe s'aggrottano sopra ed hanno un color di sangue sbiadito colà dove la ruggine manca nel bianco. (1894, 10)

In other words, this is a place where the deadly revenge an enormous slab of animated marble may take toward men is viewed as a justified response to the greedy—and yet also somehow pathetic, in terms of size disproportion—attempt of the latter to remove it from its womb-like "holy native place." Once again, Ceccardi's words suggest that marble is very much alive—to the point that it may react to the "rape" of the extraction process and crush the human culprits to death.[9]

Ceccardi's description of the quarries as "enormous candid wounds" that stand out from the rusted gray color of the surrounding mountains and his reference to the blood-like streaks of dirt that occasionally tinge the whiteness of their extension further emphasize the association of the lacerated, blown-up land with a violated, organic body.[10] Ceccardi's awareness of the interconnections between the quarry and the workers' bodies; his understanding of their disruptions and malaise in a scenario where both entities, as Alaimo puts it, are part of "power structures that have material effects" (2010, 86); and his strong sense of social justice thus emerge throughout this piece.[11] Thus as Ceccardi attempts to give a voice to the poor "little guys" (*piccoletti*) who are outsized by the surrounding nature and exploited by an authoritarian political and economic system, he transcends the site-specific situation of the quarrymen in the Apuan Alps and links these regional dynamics to the much larger horizon of the advent of global capitalist modernity:[12]

> Perhaps no lady whenever she plunges throbbing into a bathtub made out of a marble boulder from Luni has ever thought that perhaps that same boulder, as it rolled down from the peak where the Plutonian force of the Eocene had lifted it, one day got wet with the blood of the brave man who removed it from there, in a terrible baptism. And, similarly, perhaps she will never think that the pearls in the necklace she will soon wear to go out cost the life of a poor, hungry black man who drowned in the mysterious depths of the blue Ocean.
>
> E forse nessuna signora quando si tuffa palpitando in una vasca di masso lunense, ha mai pensato che forse quel masso un giorno rotolando dal picco dove la forza plutonica dell'Eocene lo aveva sollevato, si bagnò del sangue dell'audace che lo staccò, terribile battesimo, come forse non penserà mai che le perle onde si adornerà qualche momento dopo uscendo, son costate la vita ad un povero negro affamato nelle profondità misteriose dell'azzurro Oceano. (1894, 11)

Ceccardi's text deterritorializes local spatiotemporal coordinates by linking the extraction of marble and its socioenvironmental cost to the search for pearls in some far away, exotic "blue Ocean." Moreover, the deadly risks to which the workers from Lunigiana are exposed in the mountains are connected to similar dangers "a poor, hungry black man" encounters as he ventures into the depths of the sea, just as the lady's leisurely plunge in her marble bath mirrors the pearl seeker's hunger-driven dive. In this simple yet powerful image juxtaposition, Ceccardi captures the convenient, self-serving inattention and simplification that a rising Western consumer-oriented society displays toward the violence embedded in the commodities it purchases—be it a marble tub or a pearl necklace. As such, *Dai paesi dell'anarchia* not only is characterized by an early, necessarily simplified version of what Ursula Heise, more than a century later, would call a "sense of the planet"—that is, "a sense of how political, economic, technological, social, cultural, and ecological networks shape daily routines" (2008, 55)—but could also be viewed, as Rob Nixon states, as a "work of 'militant particularism,' one that discloses through that radical particularity temporal and spatial webs of violence on a vast scale" (2011, 46).

Although Ceccardi may be accused of being partial and ideologically biased, his undisputed familiarity with these places, his role as a direct witness to the events ("I have witnessed" [1894, 8; Io sono stato]), and the fact that his text was censored suggest that his writing was indeed inconveniently sensible, well documented, and obviously problematic for those who wished to continue doing business as usual. Thus this text should not be dismissed as a rhetorically constructed invention of a minor bohemian poet; rather, we should confer to it a prescient, realistic, and authoritative protoactivist voice that speaks to the human and environmental cost of marble excavation.

Crossing Geographic and Ontological Borders with Enrico Pea and Maurizio Maggiani

Among the heterogeneous adherents to the "Republic of Apua"—the circle of socialist, antiauthoritarian, and anarchist-leaning artists, intellectuals, and revolutionaries Ceccardi led for almost a decade before his death in 1919—there is the name of self-taught poet, novelist, and playwright Enrico Pea (1881–1958).[13] Born in the town of Seravezza, near Lucca, and eventually discovered by Giuseppe Ungaretti, whom he met during his emigration period in Alexandria (1896–1910),[14] Pea is considered "one of the very few actual working-class Italian modernists" who, together with the other two "Egyptians," Ungaretti and

Marinetti, "exploded the literary conventions of late nineteenth-century and early twentieth-century literary discourse in Italy" (Re 2003, 167).[15] Scholars remember Pea for his narrative work and in particular for the autobiographical trilogy that includes the novels *Moscardino, Il volto santo* (The holy face), and *Il servitore del diavolo* (The devil's servant; all three now collected in Pea 1979), in which a loquacious grandson vividly narrates the origin of his family and evokes the complex figure of his grandfather as it emerges from the background of Versilia, the Lunigiana hills and mountains between Tuscany and Liguria where the author spent his childhood. Admired by the likes of Italo Svevo, Eugenio Montale, and Italo Calvino and praised by Ezra Pound, Pea is important for my investigation here because he shares many of Ceccardo's ideological positions and makes references to native places.[16] Also, when Pea was in Egypt—experiencing an instructive, critical distancing from his Western cultural roots—he ran a marble and oil export business. What I wish to emphasize here, while also examining the contiguity among ecological concerns, new materialist theories, and posthuman positions, is Pea's "openness of mind . . . egalitarian, antiracist, antinationalist and anti-imperialist feelings" (Re 2003, 182) and, above all, his distinctive nonhierarchical, horizontal awareness of and sensibility toward people, animals, and things—the human and the nonhuman.[17]

As an example of Pea's openness of mind, consider the following passage from his *Vita in Egitto* (Life in Egypt; 1982), in which he describes one of his boat trips back to Italy. In her analysis of the text, Lucia Re observes that Pea distances himself from "pre-fascist and fascist rhetoric of racism and imperialism," and through the human-exotic animal association and the polemical reference to Darwin, he "reads through the populist demagoguery of Italian colonialism in Africa," associating "the colonial logic that victimizes Africans because they are subhuman" to the very similar one "that continues to exploit the [Italian] peasants [like himself] by forcing them to emigrate" (2003, 178–79). Pea writes,

> The deck . . . was loaded with cages containing monkeys and tortoises from the Red Sea and the forests of Eritrea. The monkeys really resembled human beings, even in their vices. During the journey, which lasted eight days, Darwin's theories that I had studied at the university for the people were confirmed by my direct observations. I spent entire days in front of the cages containing the monkeys, with whom I spoke. They were seasick like me, and like me they complained about their captivity. And toward evening like me they became melancholy, just like me they sighed when the

sun disappeared. At the end of the journey I was sad to leave the monkeys, as if they were human beings to whom I had become attached.

> Il ponte ... era ingombro di gabbie con dentro bertucce e tartarughe pure quelle provenienti dal Mar Rosso e dale foreste dell'Eritrea. Le bertucce somigliavano proprio agli uomini, anche nei vizi: durante il viaggio, che fu di otto giorni, le teorie di Darwin, di cui avevo saputo all'università popolare, dettero spago alle mie osservazioni dirette. Passavo le intere giornate davanti alle gabbie delle bertucce, con le quali parlavo. Soffrivano il mal di mare come me e come me si lamentavano della prigionia. E verso sera come me s'immalinconivano, proprio come me sospiravano allo sparir del sole. Alla fine del viaggio abbandonai le bertucce a malincuore, come se fossero state persone umane a cui mi ero affezionato. (1982, 44–45).

Passages informed by a similar egalitarian spirit and nonhierarchical tone can also be found in earlier works that are rooted in Pea's native places rather than on a ship in the middle of the Mediterranean Sea, and in this new context, they acquire additional layers of meaning. In the following pages, I explore the Versilian landscape bordering Lunigiana, with its mountains, rivers, and marble quarries that constitutes the background of the second novel in Pea's abovementioned trilogy, *Il volto santo*, which is often considered the richest and most complex in the group.

Stringing together scattered images and dreamlike visions, this fragmentary novel is pervaded by a vitalist impulse that is familiar to Italian modernism in general and a moralist tendency typical of the writers associated with the literary journal *La Voce*.[18] It immerses readers in the rural, natural, primitive world of this part of Tuscany. Anticipating some of the positions of the soon-to-beformed Strapaese (Supervillage) movement,[19] Pea viewed such an immersion as essential for human beings to achieve a full, authentic life of freedom and serenity. The text begins with a description of a cholera epidemic in Lucca and ends with one of a flood in the Tuscan town of Seravezza, a place where "everyone ... works with marble" (1979, 201; tutti ... lavorano il marmo). Between these two extremes, the young narrator (Moscardino), occasionally alternating with his grandfather's stream-of-consciousness-like accumulation of thoughts and musings, guides readers through a series of adventures. Each episode—from Moscardino's grandfather's exit from a madhouse and his spring return to a new, rejuvenated life of new encounters, love, and relationships—carries some philosophical truth or life lesson that is rich with moral implications. As these episodes also refer to situations of human-nonhuman entanglement, they offer insight into Pea's own environmental sensibility and preecological vision.

The idea that the human characters and the landscape of Lunigiana overlap and are mutually constitutive is expressed in several of Moscardino's grandfather's streams of consciousness. For example, when the grandfather, an ex-psychiatric patient, wonders about the sanity of the old director of the hospital, his self-answering process includes the following reflections:

> Men's judgment is proportional to the environment in which they live... the body takes the appearance and shape of things nearby, from the humblest to the noblest. One takes the color of the place in which one lives.... And animals are like us, neither worse nor better... if man takes the habits and the appearance of the closest animated and inanimate things: if it is all right that Bedouins crouch like their camels, and ruminate like them, it is all right that Bedouins' tents resemble a gutted camel, it is ok that a palm tree resembles a minaret, then it is also ok that the madhouse doctors become crazy... like their patients.

> Gli uomini hanno il giudizio proporzionato all'ambiente in cui vivono... il fisico va prendendo l'aspetto e la forma delle cose vicine, dalle piú umili alle piú alte. Si piglia il colore dell'ambiente in cui si vive.... E le bestie sono come noi, né peggio né meglio... se l'uomo piglia le abitudini e gli aspetti delle cose animate e inanimate piú vicine: se è giusto che i beduini stieno accoccolati, come i loro cammelli, e come loro ruminino, è giusto che le tende dei beduini rassomiglino a un cammello sventrato, è giusto che un palmizio rassomigli a un minareto, ma è anche giusto che i medici del Manicomio diventino pazzi... come i loro ammalati. (Pea 1979, 152–53)

While these remarks can be simply seen as the "crazy" grandfather's attempt to rationalize and explain to himself and his grandson a particular human behavior, they also strongly suggest his (and, in turn, Pea's) intimate understanding and joyful acceptance of a reality in which human bodies and minds, things, plants, and animals are not independent entities but rather parts of a collective network where each affects the other—a world, as he puts it, where "everything is possible" (152; tutto è possibile). Thus passages like this seem to question strictly anthropocentric interpretations that suggest that in *Il volto santo*, human beings, emblematized by Moscardino and his grandfather, are uniquely and self-sufficiently responsible for their vitalistic renewal and rejuvenation "by their own means, their own choice and initiative" (Guarnieri

1979, xi; con i propri mezzi, con una propria scelta e una propria iniziativa). Rather, Pea indicates that personal enrichment, creativity, and fulfillment can be achieved only when the subject is able to establish a nonsolipsistic, nonselfish open dialogue and relation with every object, in the widest possible meaning of the term, in the surrounding environment.[20] As Italian philosopher and zoo anthropologist Roberto Marchesini observes, "If [the human] tried to perform an autarchic ontogenesis . . . s/he could never become human. . . . As a matter of fact, it is this hybridization, the outcome of serial processes of conjugation, that makes an individual ontologically unstable . . . dependent on and fascinated by alterity, in a continuous state of creative non-balance" (2002, 63).

More specifically, I interpret the idea of unlimited freedom that the novel celebrates by fostering "a behavior free from bonds and prejudices" (Guarnieri 1979, xv; comportamento libero da vincoli e pregiudizi) and a sense of belonging to a larger universe as fundamentally unrelated to the "exaltation of a full self-control" mentioned by Guarnieri (xv; esaltazione di una piena padronanza di sé). On the contrary, I view this freedom as linked to the recognition that external factors play an essential role in the formation of the self. From this perspective, the self would depend on the loss of individual sovereignty and on an alternative exaltation that comes from the realization that being an unstable, incomplete, fragmented, and yet receptive part of the whole—far from being madness—is the foundation on which to build a more comprehensive notion of the human and to foster one's creativity. Freedom, in this sense, is the liberty to think beyond restrictions and ontological borders. The narrative path toward feeling integrated as a citizen of the larger world, in other words, is marked by an impure philosophy of hybridity and entanglement—one that is ironically delivered in the novel by an ex-madman who has actually figured it all out. The following flux of enigmatic, existential truths pronounced by Moscardino's grandfather exemplifies this situation:

> What one says for men, is valid also for the other animals, and for things. . . . Moscardino, did you ever see a dead snake? Didn't it seem to you like a root plucked up from underground? . . . Every time I bare the roots of a plant, and I expose them, I am afraid to hurt her toes. I feel that the roots are so soft under the mattock, like human muscles; and if by chance I peel some of them, I immediately feel like wrapping a bandage around them, fearful that they may bleed at any moment. I see all the things obeying the same law. Everything seems to me alive in the same way, and, if [it is true that] plants adapt

[to their environment], I don't think that the animal and the vegetable are two separate reigns. So, this truth is valid for everything. In Lucca, there are families that descend from the roots of distant trees, but they are not completely vegetable because they have dragons among their relatives. They are also kin with other minor animals, with birds; and even with the stars.

Quello che si dice per gli uomini vale anche per le altre bestie, e per le cose. . . . Hai visto mai, Moscardino, un biscio morto? Non ti sembrava una radice svelta di sottoterra? . . . Tutte le volte che scalzo una pianta, e scopro le radici, ho paura di farle male ai diti dei piedi. Sento le radici cosí morbide sotto la marra, come muscoli umani; e, se mi viene fatto di sbucciarne qualcuna mi piglia voglia subito di fasciarla, per paura che debba scianguinare da un momento all'altro. Vedo tutte le cose ubbidienti alla stessa legge. Mi par tutto vivo nello stesso modo, e non credo che sieno due regni quello animale e quello vegetale, se anche le piante si adattano. Questa verità vale dunque per ogni cosa. . . . A Lucca, vi sono famiglie che discendono dalle radici di lontani alberi, ma non sono perciò del tutto vegetali, perché hanno dei draghi per parenti. Sono imparentate anche con altri animali minori, con volatili; e perfino con le stelle. (Pea 1979, 155, 159)

A snake can "be" a root, and a root that is dug up can "be" and bleed like a wounded human finger. Everything depends on the same laws. Everything is alive and becomes and adapts to the surrounding environment. Trees, dragons, birds, and stars can all be part of the same genealogical tree because they are constituted by a similar, primal, cosmic matter. Here, Pea's amplification of the notion of the human and his representation of a fundamental "vitality of things" may recall—although in a different, less hyperbolic and more measured tone—some of the vibrant futurist positions on matter seen in chapter 1. This similarity, however, should not be surprising, especially if one considers not just that Pea and Marinetti spent time outside of Western culture in Egypt but also that they belong and contribute to a modernist milieu that has repeatedly confronted the shifting ontological extension of the subject and its relation with the natural world in an atmosphere of radical changes, crises, and alienation. In addition, Pea's overall ambition in *Il volto santo* to represent the impossibility for human beings to be a single thing, tied to just one role during their existence (a sinner or a saint, a devilish madman or an angel, a human being or an animal), strongly suggests a Pirandellian presence in his novel. In particular, we may think of Luigi Pirandello's *One, No One, and One Hundred Thousand* (1992), with which it shares the theme of madness. There is even a peculiar similarity between the names of the protagonists: Moscardino and Moscarda.[21]

According to Luca Somigli's interpretation of Pirandello's novel, the main character "finds in the natural world that fragmentation that characterizes his own life" (2016, 142). This is evidenced when Moscarda states, "I am alive and I do not conclude. Life does not conclude. And life knows nothing of names. This tree, tremulous pulse of new leaves. I am this tree. Tree, cloud; tomorrow book or wind: the book I read, the wind I drink. All outside, wandering. . . . I die at every instant, and I am reborn, new and without memories: live and whole, no longer inside myself, but in everything outside" (Pirandello 1992, 160). Likewise, in Pea's *Il volto santo*, there is a similar movement from the outside toward the inside, from the environment toward the human body (and vice versa), and therefore a parallel extension of the borders of the human in nature. However, I read this tendency as an acceptance of and availability to be and become something else, regardless of the fact that this "else" may be perceived as less or more than one's original status. In other words, the notion of fragmentation in Pea's work does not necessarily carry a negative connotation, and nature is expected not merely to "recompose" but to mirror, complement, and cooperate with human beings.

According to the imaginative, eminently ecological dynamics displayed throughout Pea's trilogy, "measure is what counts. . . . If you go beyond a certain measure, you become a different thing from what you were before" (1979, 157; la misura è quella che conta. . . . Se oltrepassi una certa misura, sei un'altra cosa da quella che eri prima).[22] The introduction of this notion of measure, which appears most frequently in *Il volto santo* to indicate the existential benefits of displaying an attitude of moderation that urges human beings not to "exceed, to demand from things and from ourselves more than it is right, more than it befits each man" (di trasmodare, di esigere dalle cose e da noi piú di quello che sia giusto, piú di quello che a ogni uomo si convenga) and to reject "any will to excel . . . to impose oneself, to overpower others" (Guarnieri 1979, xiv; ogni volontà di eccellere . . . di imporci, di sopraffare gli altri), allows me to return to the other reason I chose to include Pea's novel in this discussion—namely, its reference to a destructive inundation in the marble town of Seravezza, one foreshadowed by Moscardino's grandfather's remark that "a river generates a flood when water goes beyond the ramparts fit to restrain it" (Pea 1979, 157; un fiume genera un'alluvione, quando l'acqua oltrepassa i bastioni che le erano stati misurati). By linking this catastrophe to the reevocation of Moscardino's father's death and the loss of his family's home on the one hand and to the places and practices of marble extraction and manufacturing on the other, this episode dramatizes and exemplifies, in both existential and environmental terms, the ultimate consequences of a lack of measure among human beings.

To put it differently, the conclusion of the novel suggests that the rejection of abusive and arrogantly self-centered behaviors not only should characterize human beings' relationship with one another but could also include an "other" that is not necessarily human. In this light, Moscardino's sensation of renewed fear when he remembers how a block of marble was pulled away from the mountain to be transported to the coast is particularly revealing:

> You are frightened when you see a cart loaded with a boulder of white marble, still bleeding with the dirt from the mountain: marked with numbers and stamps, and tied with the chains of a three-mast boat . . . when it bites the gravel . . . and above it . . . is the cart's driver balancing himself, as if he were an evil demon: and he yells, strikes and spurs the pulling oxen. You would think it is a procession of monsters that are hauling evil to hell. My father's family used to build these transport machines.
>
> Ti fa paura, quando vedi una carretta con in groppa un masso di bianco marmo, insanguinato ancora della terra della montagna: segnato con numeri e con marche, e incatenato con le catene dei barchi a tre alberi . . . quando morde la ghiaia . . . e sopra . . . sta il carratore in bilico, come un cattivo demonio: e sbraita, percuote e punge i bovi in tiro, ti pare una processione di mostri che trascinino il male all'inferno. La famiglia di mio padre costruiva queste macchine da trasporto. (Pea 1979, 199)

The power of this description resides not only in the anthropomorphism of the bleeding, marked, and chained marble boulder and of the earthly mother from which it has been violently separated but also, widening our perspective beyond the local, in its allusions to violation, enslavement, and relocation far from home—that is to say, a situation for which Moscardino's family, the people working with marble, and the people who keep blindly reducing things (human or nonhuman) to commodities are morally responsible. This suggestion is corroborated by Pea's reference to the global destinations of marble and, implicitly, to Moscardino's awareness of the dynamics of the modern market and commercial exchanges. Those who work the stone in their shops, Moscardino observes,

> chisel the ashlar-work of wealthy homes in overseas cities, their stairs, and the lintels of churches built by Christian love in Tierra del Fuego, in Africa, or in regions unexplored by Science.
>
> scalpellano i bozzati delle case ricche per le città di oltremare, e le scale, e gli architravi delle chiese che l'amore cristiano edifica nella terra del fuoco, o in Affrica, o nelle regioni inesplorate della Scienza. (Pea 1979, 200)

In other words, especially if read in conjunction with Re's remarks about Pea's "critical understanding" (2003, 175) of the colonial dynamics that he experienced in Egypt and the earlier passages about the bleeding marble block and the captive monkeys on the ship, the sense of pride that this list of human achievements may inspire is tempered and problematized by the actual cost of the commercialization of marble and its derivatives.

In fact, some of the socioenvironmental repercussions of marble excavation practices—deforestation, mountaintop removal, mining dumps, hydrogeological instability, groundwater pollution, and a general territorial impoverishment that penalizes locals while enriching foreign corporations—are foreshadowed in Pea's depiction of the flood at the end of the novel. In the following passage, the effects of men's ecological violence and lack of measure is directly related to the material overflowing from the rivers crossing Seravezza:

> Two rivers, already swollen with a dense and reddish foam: water and soil from the quarries. It was as if the mountains were bleeding, as if a massacre had happened in altitude, and the thunders and lightnings were the furies in one of those lethal battles called:—Fire and sword. Each furrow on the mountains had changed into a canal, and all these canals carried their deadly contribution into the three rivers. Birds caught in their sleep, ducks, dogs, tree trunks floated in the streets. The bridges were choking and collapsing. The water overflowed the riverbanks and was visiting the stalls. The sheep and the smaller animals were the easiest prey. One could hear the neighing of horses, the bellowing of oxen who did not resign themselves to die. . . . While my home built over the sawmill on the river was collapsing, a Christ on the cross was seen floating toward the sea.
>
> Due fiumi, già gonfi di sanguaccio denso e schiumoso: acqua e terriccio di cava. Come se i monti sanguinassero. Come se sulle alture fosse avvenuta una strage, e i rombi e i fulmini fossero le furie di una di quelle battaglie di sterminio che si chiamano:—Ferro e

> fuoco. Ogni solco sui monti era diventato canale, e tutti questi canali portavano nei tre fiumi il loro contributo di morte. Uccelli còlti nel sonno, anatre, cani, tronchi d'albero, galleggiavano nelle strade. I ponti strozzati e franati. L'acqua straripata dalle spallette dei fiumi visitava le stalle. Le pecore e gli animali piccoli furono le prede più facili. Si sentivano i nitriti dei cavalli, il muggito dei bovi che non si rassegnavano a morire. . . . Cristo in croce fu visto galleggiare verso il mare, mentre franava la mia casa costruita sopra la segheria nel fiume. (Pea 1979, 203)

While it may be redundant to underscore the militant language Pea uses to describe the violent, conflictual events in the mountains, it may be less so to observe the ambiguous duplicity of this violence. For example, the storm that naturally swells the rivers and erodes the soil is undeniably related to the artificial violence brought into this territory, where the real slaughter is the one caused by the blasts and explosions perforating the bleeding body of the mountains, the debris clogging the rivers and making them flood.

The passage from *Il volto santo*, with its list of bleeding, floating, crumbling, and screaming human and nonhuman bodies, could describe any one among the floods that in recent years, due to the disruptive effects of modern extraction technologies, have devastated several districts in the Apuane—the city of Carrara in particular. This can only emphasize Pea's environmental sensibility and the pedagogical role his narrative assumes as both a cautionary tale and a helpful means to reimagine these places in different ways.[23]

The numerous marble quarries in the Garfagnana region of the Apuan Alps and the valley and district around the Pania della Croce mountain provide the scenery for Maurizio Maggiani's *Meccanica celeste* (Heavenly mechanics; 2010). This novel combines traditional storytelling with anthropological and moral insights and manages, once again, to connect local events to a wider context. It is not by chance that its title refers to the cosmic laws governing the whole universe or that, as Giorgio Cattaneo writes, here "Garfagnana can be the world, a harsh and friendly world that adopts you and to which you surrender . . . among private and collective memories" (2010). However, the work's relevance to our discussion goes beyond its geographical background.

Alternating between the past and the present, the passages describing the life of the workers in the quarries show a deep, personal knowledge of the Apuan territory and insert themselves into a marble-influenced tradition and an aesthetic that is aware of the permeable borders between human beings and the land they inhabit. Maggiani also alludes to the violent, predatory dynamics and increasing lack of measure that continue to affect and operate in this landscape in current times. For my limited objectives and for the reader here, there is no better way to read "text and world . . . as 'circulating references'" (Iovino 2014, 10) and thus to enter the rich, fictional universe of *Meccanica celeste* than to put ourselves in the shoes of "il Vittorio," the old stone breaker who appears in the epigraph to this chapter. Like many other memorable characters in this epic narrative, Vittorio embodies a premodern, magical way of relating to the local landscape, which he sees as an assemblage of plants, animals, minerals, and human beings.

But let us make a couple of clarifications before meeting him. Given the goals of this section, it should not be necessary to contextualize Maggiani's work within the contemporary Italian cultural scene or engage in a discussion about literary genre, form, or style. However, I should point out that my use of the term *epic* to qualify this novel refers to and dialogues with the recent debates about the so-called New Italian Epic (NIE). This label was introduced in 2008 by the writer Wu Ming 1 (one of the members of the writers' collective Wu Ming) to indicate a series of literary works written in Italy between 1993 and 2008 that share some stylistic, thematic, and allegorical tendencies. While *Meccanica celeste* does not fall within the indicated temporal arc, it displays at least some of the features that would justify its inclusion among the NIE novels. In addition to its realistic register, it combines the personal, the historical, and the mythical, and similar to other NIE works, it "tells about . . . the lacerations, the chaotic diverging and becoming, the de-territorializations and re-territorializations taking place on the weakened body of an imploding, racist and livid country" (Wu Ming 1 2008, 11).

I wish to underline here the presence of what Wu Ming calls an "ecocentric gaze." This gaze, which allows one to simultaneously "see the world from outside and see oneself from outside as part of the continuum of the world" (29), implies and conveys a well-defined ethical position. Following these ecocentric lines in the text, Maggiani's landscape is represented as an entity endowed with semiotic and agential qualities that are equal to those possessed by the people who inhabit it. As the narrator puts it,

> Every plant had a name and the schoolchildren knew them all; each of them had a soul and the children spoke with each one. . . . A boy . . . had chosen as his father

a chestnut tree named Beniamino, because his real dad had died during the war and he needed another one.... Chestnuts are good and one should absolutely not be afraid of them; rather, if you get to know them, they always give a hand whenever needed ... the boy's name was Mirto.

Ogni pianta aveva un nome e ragazzi della scuola li conoscevano tutti; ognuna aveva uno spirito e i ragazzi parlavano con ciascuno.... Un ragazzino ... aveva preso per padre un castagno di nome Beniamino, perchè aveva perso il suo primo nella guerra e aveva bisogno di un altro.... I castagni sono buoni e non c'è da averne minimamente paura; piuttosto, a volerli conoscere, danno sempre una mano quando ce n'è bisogno ... il ragazzino si chiamava Mirto. (Maggiani 2010, 64, 109)

This passage should remind readers of similar depictions in Ceccardi's and Pea's texts of a literary and geographical universe of codependency where making rigid distinctions among the realms of the human, animal, vegetable, or mineral does not make sense. In a place where a chestnut tree acquires anthropomorphic features and assumes the role of a surrogate father and helper and a child is named after a typical Mediterranean plant, "il Vittorio" can nonchalantly "speak to the stones like Omo Nudo speaks to his pigs, Marta used to speak to her heifers, and Malvina spoke to the doorway of Aristo's home" (264; parla[re] ai sassi come l'Omo Nudo parla ai suoi maiali, la Marta parlava alle sue giovenche, la Malvina al portone della casa dell'Aristo).

One implication of this intersubjective and "hetero-referred" (Marchesini 2014, xv) dialogue with the nonhuman is that those human characters who are able and willing to practice it—the term *human* here indicating their status as heirs (*eredi*) and descendants (*discendenti*) of the "seed of Adam . . . masters of the whole of creation" (Maggiani 2010, 163; seme d'Adamo ... padroni di tutto l'universo creato)—necessarily undergo a substantial change, becoming more open, inclusive, and hybrid. Figures such as "il Vittorio" or, even more explicitly, the "Omo Nudo" with his natural, nonobscene nakedness and synchrony with the rhythms of nature, nostalgically embody a less rigid notion of humanity. Therefore, they are living proof that

Adam's seed dispersed among the stones squared by his descendants ... it bastardized into something inhuman in which he [Omo Nudo] ran into while squaring stones and putting them one on top of the other.

> il seme di Adamo si è disperso tra le pietre squadrate dei suoi stessi discendenti ... si è imbastardito in qualcosa di disumano in cui si è imbattuto squadrando le pietre e mettendole una sull'altra. (163)

At first sight, words such as *bastardized* and *inhuman* seem to possess a negative connotation and indicate a regression. However, in the paragraphs that follow, the intradiegetic narrator characterizes this sequence of the bastardization, inhumanization, and dispersion of humanity's seed among the stones as a series of lucky events, situating them in opposition to the truly regressive forms into which modern humanity can paradoxically evolve on its allegedly progressive path:

> We don't even preserve Adam's memory, and this is fortunate ... we are content with our limited and meager lordship, suited to our condition. But the Lombards who come here with their pickup trucks, wearing camouflage gear and carbon-fiber jackets, carrying with them military rifles to shoot short-range at Marta's heifers, they don't even dream of legislating on something. They are just robbers who elect themselves masters of the whole universe, created and not. That is to say, of everything they can put their hands on.

> Non conserviamo neppure il ricordo di Adamo, ed è una fortuna ... ci accontentiamo di una signoria ristretta e meschina, adatta al nostro rango. Ma i lombardi che arrivano con i loro pick-up abbelliti di cerate mimetiche e le casacche in fibra di carbonio ornate di carabine belliche, per sparare da duecento metri alle manzette della Marta, loro non si sognano neppure di legiferare su qualcosa. Loro sono solo rapinatori che si eleggono padroni, padroni di tutto l'universo creato e non creato. Ovvero di tutto quello su cui riescono a metter le mani. (163)

In a passage that alternates between the local and the cosmic, the narrator reveals details about his ontological vision and ecological sensibility. In short, Maggiani argues that if being a modern, evolved human being means avoiding any responsibilities or sense of measure and behaving arrogantly, violently, and rapaciously toward the whole universe, then it is better to forget about our supposedly higher nature and origin. Situated at the behavioral antipodes of such Northern Lombards, "il Vittorio" instead shows respect, consideration, and empathy even for the fragments of discarded marble that he recovers along the river in order to make his living and that, in turn, define his own human-mineral self; Marchesini would likely say that the minerals themselves collaborate in the making and realization of this character's body. By carefully "listening to what they have to tell him" (Maggiani 2010, 264; stare a sentire cosa hanno da dirgli), "il Vittorio" embodies what the narrator calls a "limited and meager lordship" (163; signoria ristretta e meschina) on the landscape he inhabits and coconstitutes. It is thus tempting to think of his character as the latest listener and collector of the tales told by the mountains and the rivers mentioned earlier in Ceccardi's poem or to imagine him as a close relative of Moscardino's grandfather. At the same time, "il Vittorio" serves as a witness to the physical and cultural transformations that have been affecting this territory and as an emblem of resistance to the many assaults brought over the years against these particular (but also universal) places and people, from the Este dynasty domination in the sixteenth century, to the struggle with Nazi fascists during World War II, and to modern traumas linked to marble extraction, industrialization, and land-grabbing practices. These evils, however—as the narrator suggests toward the conclusion as he comments on M. P. Shiel's 1901 science-fiction novel *The Purple Cloud*—should never prevent us from abandoning the hope of envisioning a better future for humanity. On the contrary, the individuation and narration of such traumas and evils is the first step toward restoring "the richness, diversity and complexity of those inner landscapes of the mind . . . which are threatened by impoverishment from an increasingly over economized, standardized, and depersonalized contemporary world" (Zapf 2010a, 10).

At this point, what else may the stones and marbles of the Apuane tell "il Vittorio" as he wanders through this geographical and literary landscape? Are there other ways to explore the narrative power of marble—its influence on the literature and cultural production of this area—and to investigate further some of the socioecological implications of the bodily and material entanglements that the texts considered so far have begun to illuminate? In order to answer these questions and, building on Hubert Zapf's remarks, to emphasize the role of storytelling and of "the culture-nature relationship as . . . site[s] where ecological concerns and the ethical self-reflection of the human species are brought together" (2010a, 12), I consider the two contemporary nonfiction accounts mentioned earlier that may help seal this discussion. As Marco Rovelli's *Il contro in testa* and Giulio Milani's *La terra bianca* illuminate the close relationship between marble extraction and other scarring forms of industrial exploitation of this territory, they also retrace the mutual connections among the disruptive forms of violence (political, physical, cultural, environmental) that have long marked the history of the Apuan Alps. For continuity's sake, let us also observe that both these writers can be viewed as heirs and contemporary spokespersons of Ceccardi's and Pea's anarchic tendencies, especially with regard to their environmental and ecological sensibilities. As Milani states, "The anarchist component in which I recognize myself is ecological and nonviolent" (2015, 177; La componente anarchica in cui mi riconosco io è ecologista e non violenta).

Toxic Autobiographies: Marco Rovelli's *Il contro in testa* and Giulio Milani's *La terra bianca*

As Carlo Mazza Galanti puts it, Marco Rovelli's *Il contro in testa* is many things at once: "[It is] a poetic text about the exceptional political story of this region. A travel book in a territory filled with stories which moves more vertically through time and memory . . . than horizontally in space. An autobiographical sketch of a whole generation" (2012). With a title that alludes to the natural, geological resistance of marble to being cut against the grain and that also evokes the historically rebellious and independent nature of the population in the marble region of the Apuan Alps, this book connects the mineral and the human world and introduces a scenario where, once more, men and things belong to the same ontological horizon. As the first-person narrator intertwines quotes from local interlocutors and his personal history with the much longer history of the quarries, he suggests that the latter's privatization among multinational entities played a role in the emergence of sentiments of resistance and acts of protest in the territory. These include both those expressed by the mountains themselves and those voiced by the anarchists, the antifascist partisans, and the citizens concerned about their degrading environment. By showing how both the population and the land have been protagonists in parallel stories of violence, exploitation, poverty, displacement, and degradation over the years, I argue that *Il contro in testa* not only echoes and expands the message of denunciation and social justice traceable to Ceccardi's *Dai paesi dell'anarchia* but inserts itself into the Italian version of a genre of recent environmental writing known as "toxic autobiography."

Originally described by Richard Newman as a narrative form that "meditates on the personal, political, and historical meanings of the hazardous waste grid that developed in the United States following World War II" and that "flows from a deep sense of crisis among marginalized groups of people . . . that feel trapped in poisoned landscapes well beyond mainstream concern," toxic biography's "second wave" is "coming from professional or artistic backgrounds (as reporters, writers, and scientists) . . . [who] produced more polished and stylized narratives—almost novelistic in their portrayal of toxic experiences" (2012, 22, 37). Rovelli's book does not address a specific chemical spill or a single case of toxic pollution like those Newman discusses. However, Rovelli's role as a writer, teacher, and musician, together with the hybrid nature of a text that is part investigative journalism and part autobiographic novel, should leave few doubts about his text's inclusion in this genre. It is clear that Rovelli desires to raise his readers' political awareness about the progressive degradation and crisis of the Apuan Alps territory due to a mass industrialization that, as he notes, was literally "born out of marble" (2012, 118; nata dal marmo).

As the narrator begins his journey "in this land of ghosts" (10; terra dei fantasmi), he evokes past literary and political figures who resisted the various forms of violence brought onto the land and combines these references with contemporary figures from the local community (retired quarry workers, political activists) and excerpts from his own biography. In this way, Rovelli presents a survey of the sociopolitical and environmental history of the territory. The narrative also shows how violence "is disseminated as a language and as a system of signs in these grounds and bodies" (Iovino 2016a, 139). In this light, distant and apparently unrelated remarks become poignantly meaningful when considered together and perfectly illustrate the dissemination of violence in both the earth and human bodies:

> Should we remember that this Apuan land is one of slaughters, where hundreds of civilians were massacred? . . . Up from the Vergheto one sees a plane mountain, gutted by the quarries.
>
> Terra d'eccidi, questa apuana, serve ricordarlo? . . . Centinaia di civili massacrati dal Vergheto si vede una montagna piallata, sventrata dalle cave. (Rovelli 2012, 9)
>
> One feels a sense of dismay, a shiver in seeing this violence. This unusual profile, this tooth that stands out lonely in a hollow mouth-like cavity has, in its awfulness, a perverse fascination.

> C'è un moto di sgomento, un brivido, nel vedere questa violenza. Ha un fascino perverso, nel suo tremendo, questo profilo innaturale, questo dente che spicca, solitario, in una vuota cavità buccale. (41–42)

The massacre of innocent civilians perpetrated by the Nazi invaders during World War II is mirrored in the carnage perpetrated toward the land itself, where the profile of a gutted mountain strongly resembles a tooth in the gaping mouth of a corpse. The narrator's observations convey the difficulty of delineating a clear-cut line between human and nonhuman bodies and invite readers to interpret the parallel violence that was done to both people and landscapes, inner and outer worlds.

As it turned out, the Nazi armies were not the only foreign invaders this territory was going to face. As Rovelli makes clear, the fascist regime's creation of the Zona Industriale Apuana (ZIA; Apuan Industrial Zone) in 1938, between the cities of Massa and Carrara, was a response to a crisis in marble exportation caused by the regime's own autarchic policies, and this was only another, softer form of invasion that allowed big industries from Northern Italy to begin conquering and colonizing this land and exploiting cheap labor:

> It was because of the crisis in the marble business.... The big Northern industrialists descended to colonize the plain between Massa and Carrara, having a large and cheap workforce available [for their] steel, mechanical, chemical industries.
>
> Fu per la crisi del marmo.... I grandi industriali del Nord scesero a colonizzare la piana tra Massa e Carrara, avendo a disposizione pure una grande forza lavoro a basso costo. Siderurgia, meccanica, chimica. (Rovelli 2012, 118)

Over the years, this area came to host factories owned by both major Italian corporations, such as Ente Nazionale Idrocarburi (ENI; Italian Hydrocarbon Corporation) and Montedison (Syndial, Rumianca, Farmoplant), and foreign multinationals.[24] Rovelli's first-person evocation of the series of subsequent environmental disasters—culminating with the 1988 explosion of the Farmoplant chemical plant (a.k.a. the "Italian Chernobyl")—and of the eventually successful efforts of local grassroots organizations to fight against the pollution of their land and bodies justifies his text's inclusion in the category of "toxic autobiography":

At that time, the manifestations I attended were those against the chemical plant, its dangerousness, the exponentially increased tumors. The struggle against Farmoplant was a formative one for my generation. . . . The plant is 99.999% safe, we were told. Sure: on July 17, 1988, there was the ultimate accident. The drums containing Rogor, the most harmful pesticide, which Farmoplant sold as the purest in the universe, exploded. . . . Whoever was able to do so escaped from Massa on that day of real panic. Those who stayed went to protest under the Ducal palace and were tackled and clubbed by the police. But it was the end for Farmoplant. And yet a wasteland has remained. . . . The liberally spreaded poisons have remained, together with a petrified land. Nobody paid damages to the polluted people.

Le manifestazioni cui partecipavo, all'epoca, erano quelle contro il polo chimico, la sua pericolosità, i tumori cresciuti esponenzialmente. La lotta contro la Farmoplant è stata una lotta formativa per quelli della mia generazione. . . . Fabbrica sicura al 99, 999%, era scritto. Eccome: il 17 luglio 1988 l'incidente definitivo. Esplodono i fusti di Rogor, il pesticida più nocivo che la Farmoplant spacciava come il più puro dell'universo. . . . Chi poté scappò da Massa, quel giorno, fu vero panico. Chi restò andò sotto il palazzo ducale a protestare e venne caricato e manganellato dalla polizia. Ma per la Farmoplant fu la fine. È rimasta però la terra desolata. . . . Sono rimasti i veleni cosparsi a piene mani, è rimasta una terra pietrificata. Nessuno ha pagato i danni al popolo inquinato. (113–15)

By siding with and giving voice to the "colonized" locals and expressing the tension between the ecological threats posed by technical modernity and its promise of improving the economy and people's lives, Rovelli's activist writing helps "uncover [a] poisoned past and remap[s] the meaning of the nation's toxic grid" (Newman 2012, 43). Along the path of other Italian chroniclers of the "happy marriage between industry and poor country" (Revelli, quoted in Iovino 2016a, 139), such as the quasi-homonymous Nuto Revelli in Piedmont's similarly degraded Val Bormida region, *Il contro in testa* reminds us of the unavoidable transcorporeality of these landscapes, where the terrain is poisoned and "petrified" and the people are "polluted." And similar to the other texts we have considered in this book, Rovelli's work emblematizes the redemptive, reintegrative power of narrative to bring readers to socioenvironmental consciousness and to imagine a future that avoids repeating past mistakes.

Moving on from Rovelli's text, the main goals of Giulio Milani's *La terra bianca*, as the first-person narrator states, are "to reclaim. To put in the right and, if possible, to forget" (2015, 6; bonificare. Rimettere nel giusto e, se possibile, dimenticare). These objectives may already justify the inclusion of this additional example of Italian "toxic autobiography" and activist writing among the narratives of redemption, reintegration, and resistance considered here. Similar to Rovelli's book, *La terra bianca* is written by a local, politically involved, and socioenvironmentally conscious author who is familiar with the history and geography of the Apuan territory. Like his predecessor, he tries to come to terms with the various forms of devastation and health problems in this area through the enlightening and healing power of storytelling after these collective traumas.

Heeding Rovelli's text, Milani's intertwines personal memories, stories of the local partisan resistance, reports of factual events, witness accounts and interviews, and imaginative writing to weave together a narrative that reads like a mystery novel and attempts to investigate "a problematic, often-unacknowledged past that valorized industrial and high-tech development above human and environmental safety" (Newman 2012, 44). Quite contradictorily, although one of Milani's declared objectives is "to forget," it becomes evident that he is interested in remembering facts, reconstructing events, interpreting the language of people and landscapes, and sharing his indignation and findings with his readers. Specifically, his story tries to make sense of the "Great Trauma" (Milani 2015, 6; Grande Trauma), Milani's term for the explosion of the Farmoplant chemical plant on July 17, 1988, and of its far-reaching impact on both humans and the environment. As the author puts it in a revealing metaliterary reflection about the truth-searching role of the writer in a society that has stopped assigning any ethical or public function to literature, he perceives himself as a

> detective trailing a crime, or a trauma, in search for those responsible. In order to do that, he paradoxically started from a known finale—the traumatic event—and then pieced together its sequence, its past developments, situating his fantastical hypotheses between the future perfect and the past conditional. So it happened that a fact B that occurred today found its foreshadowing in a fact A that took place even earlier in time.

> detective sulle tracce di un crimine, o di un trauma, di cui doveva rintracciare i responsabili. Per farlo, cominciava paradossalmente da un finale noto—l'evento

traumatico—e ne ricostruiva la sequenza, gli sviluppi passati, lanciando le proprie
ipotesi romanzesche a metà tra il futuro anteriore e il condizionale composto. Così
capitava che un fatto B, successo oggi, trovasse la propria prefigurazione in un fatto
A accaduto anche molto avanti nel tempo. (155)[25]

The idea that a certain "fact B" happening today can, and actually needs to be, explained by looking at an earlier "fact A" confirms the connection Rovelli suggests exists between the partisan war against the Nazi fascists and the struggle against more recent forms of socioenvironmental violence and oppression[26] and the link between the older marble industry and the later chemical hub in the ZIA—or, as he puts it, between "chemistry and calcium carbonate on the one hand, and ecomafia and turbo-capitalism on the other" (60; chimica e carbonato di calcio da una parte, ecomafia e turbocapitalismo dall'altra).[27] The same idea can be considered in light of Rob Nixon's notion of "slow violence"—that is to say, "a violence that occurs gradually and out of sight, a violence of delayed destruction that is dispersed across time and space, an attritional violence that is typically not viewed as violence at all" (2011, 2).

Despite the fact that Milani's story begins with a very visible explosion, its development makes clear how this catastrophe has remained invisible to the public:

So many people, workers and residents, died and continue to die in this "land of fires" that tops the charts with regards to blood and respiratory diseases but that is still basically unknown to the national public.

Quante le persone, lavoratori e residenti, che sono morte e continuano a morire in questa "terra dei fuochi" da primato per malattie del sangue e respiratorie, ma ancora pressoché sconosciuta all'opinione pubblica nazionale. (2015, 7)

His narrative also reveals how the explosion's causes, effects, and repercussions are fully traceable only by looking backward and forward in time and space and thus considering a "range of temporal [and spatial] scales" (Nixon 2011, 2).

Milani's speculative narration thus manages to make visible the particular and yet reproducible history of violence to landscapes and bodies that, in this case, begins with marble extraction in the Apuan Alps, continues with the catastrophic chemical transformation of the industry and environment, and then progressively and translocally "disperse[s] across time and space" (Nixon

2011, 2) into other areas of Italy and wherever reckless industrial practices and policies of waste management affect the health of local inhabitants. Milani's description of the conditions at the ex–Rumianca poisons factory in the late seventies and eighties gives readers an idea of the risks workers were exposed to and of the transcorporeal, material entanglements that were taking place:

> The powder [a by-product of the citric acid used to treat some marble slabs] gave a blue color to the workers and transformed them into a surreal human landscape: organisms made of flesh became in turn a sort of factory and mill, where pipes, blast furnaces, tanks cohabited with mallets, with grease pillows. Scalding steam arose from them. Dark or light fires blazed up. Open sky streams dragged the waste with their bitterness. Other chemical reactions that sent out pestilential smells occurred at once.
>
> La polverina colorava i lavoratori di blu e li trasformava in un paesaggio umano surreale: organismi di carne diventavano a loro volta delle specie di fabbriche, mulini e frantoi, dove tubature, altiforni, vasche coabitavano coi magli, coi cuscini di grasso. Il vapore vi sorgeva, bollente. Fuochi cupi o chiari avvampavano. Ruscelli a cielo aperto trascinavano scorie col fiele. Avvenivano subito altre reazioni chimiche, che emanavano odori pestilenziali. (2015, 90–91)

In this sense, Tuscany and the Apuan region become much closer than they appear on a map not only to Campania but also to Piedmont's Susa and Bormida valleys, Puglia's Brindisi and Taranto, and Sardinia's Sulcis region:

> This territory or, better, the chemical plants active in this territory are seriously, criminally responsible for the tragic situation in the "Land of fires." . . . But the "Land of fires" . . . is here too, even since earlier times.
>
> Questo territorio, o meglio, le industrie chimiche che hanno operato in questo territorio, hanno gravi e criminali responsabilità nella situazione tragica della "Terra dei fuochi." . . . Ma la "Terra dei fuochi" . . . è anche qui, da noi e da ben prima. (110–11)

In other words, even Tuscany and the Apuan Alps have become one with all the places belonging to "the Italian industrial metanarrative" (Iovino 2016a, 140) and to that of a constantly shifting Global South that has been suffering from pollution and environmental degradation. But this is also where attempts at a "memory recovery" (Milani 2015, 128; recupero della memoria) and various forms of restorative resistance, including a narrative one, have managed to rise. Ultimately, Rovelli's and Milani's books call attention both to marble's active responsibility in shaping and supporting local histories, lives, and landscapes and to its much wider (more or less direct) transversal entanglement with universal practices and behaviors that undermine or erase such support. If, on the one hand, marble excavation in the Apuan Alps—in a way similar to coal mining in West Virginia—can be viewed as a lifeline for indigenous people, on the other, it also communicates the destruction and deterioration of a geographic and human landscape that is increasingly colonized by the logic of privatization of profits and socialization of costs.

Barry Lopez once observed that "no voice, by merely telling a story, can cause the poisonous wastes that saturate some parts of the land to decompose, to evaporate," and yet he also noted that "a testament of minor voices can clear away an ignorance of any place, can inform us of its special qualities" (1989, 60). I am not sure if the various voices we have heard so far—from Ceccardi's, to Milani's, and to the storied landscape of Apuan marble itself—should be qualified as minor in the traditional (and Lopez's) meaning of the word. However, perhaps they come closer to Guattari and Deleuze's definition of *minor* in the sense that they all seem to intrinsically possess a political nature and a collective, revolutionary, spatial value—deterritorializing one terrain as they map a different one in order to "express another, potential community, to force the means for another consciousness and another sensibility" (1983, 17). Thus despite the fact that, at least occasionally, "the stories stones tell can be dispiriting" (Cohen 2015, 63), learning to listen to them and to consider marble as one among the many material companions in our lives rather than as a simple "self-evident asset, inert commodity, or resource for extraction" (41) may be the best way to reach such an urgently needed ecological consciousness and sensibility.

4

Steel and Asbestos

Stories of Toxic Lands and Bodies in Tuscany and Beyond

Barry Lopez's observation about the potential of a "minor" storytelling practice to defuse ignorance, increase environmental awareness, function as an instrument of resistance and change, and, as Hubert Zapf writes, "reshape the ecocultural imaginary" (2016, 50) strongly resonates in the present chapter. As the title "Steel and Asbestos" suggests, it subverts traditional touristic depictions of Tuscany by initially investigating how two recent narratives have responded to the ecobiological crises linked to the perhaps unexpected presence and implications, in this stereotypically beautiful Italian region, of major industrial sites and by continuing to think from an ecomaterialist perspective about the interactions of bodily ecologies, landscapes, and socioeconomic structures. By focusing on substances like steel and asbestos, our ongoing investigation of the proximities of human beings and more-than-human environments and of various forms of agentic materiality underscores the by-products of risk and toxicity that often emerge in these circumstances.

Located at the northern tip of Tuscany's Maremma, the industrial port town of Piombino and its coastal surroundings—dotted with a large steel plant, shipyards, and refineries—provide the general background for many of the works considered in this chapter. Maremma is a mostly rural, mineral-rich area that already in the fifties was at the center of one of the earliest *ante-litteram* Italian socioenvironmental inquiries, *I minatori della Maremma* (The Maremma miners; 1956) by Carlo Cassola and Luciano Bianciardi. At the same time, like other sites in Italy and elsewhere that share many of its characteristics and negotiate regularly with the manufacturing and the environmental impact of steel and asbestos (e.g., Genova, Taranto, Terni, Bagnoli, Casale Monferrato), this place represents a convenient pretext and entry point for a discussion of the wider effects and toxic legacy of these substances.

Consistent with the methodology and conceptual framework that inform this whole project, my approach in this chapter mostly draws upon, among others, the work of Stacy Alaimo, who has discussed the transcorporeal circulations between human-nonhuman ecologies and the notion of "material memory" (2010, 2); Richard Newman's related insights on the genre of toxic autobiography as a means to "meditate on the personal, political, and historical meanings of . . . hazardous waste grid[s]" (2012, 22); and Marco Armiero and Ilenia Iengo's (2017) project on toxic bios.

"A Life of Metal": An Ecocritical Reading of Silvia Avallone's *Swimming to Elba*

> As Alessio walked he crushed nettles and chunks of refractory brick underfoot. Metal saturated the ground and his skin.
>
> Alessio calpestava ortiche e resti di mattoni refrattari. Il metallo saturava il terreno e la sua pelle.
>
> —Silvia Avallone, *Swimming to Elba*
>
> And human metalworkers are themselves emergent effects of the vital materiality they work.
>
> —Jane Bennett, *Vibrant Matter: A Political Ecology of Things*

According to its official tourist website, the section of Tuscan shoreline overlooking the Tyrrhenian Sea branded as the Etruscan Coast extends approximately one hundred kilometers, from Livorno in the north to Piombino in the south. Directly to the west lie three of the islands forming the Tuscan Archipelago—Elba, Capraia, and Gorgona. The other four—Pianosa, Giglio, Montecristo, and Giannutri—are scattered a bit farther south. To the east of the coast, in its adjacent hinterland, villages such as Bolgheri and Castagneto Carducci enhance the Etruscan aura of beauty, civilized refinement, layered history, culture, and literariness (Costa degli Etruschi, n.d.). Besides Giosuè Carducci—the bard of unified Italy, who celebrated these places in his poetry in the late nineteenth century—Gabriele D'Annunzio spent at least some of his "inimitable life" in his cliff-side Villa Godilonda near Quercianella. A little later, yet another famous poet, Giorgio Caproni, wrote memorable lyrics about Livorno (his birthplace) and his beloved Tyrrhenian Sea. And finally, in 1962, film director Dino Risi shot some crucial scenes of his classic *Il Sorpasso* (*The Easy Life*) on the coastal road near the fashionable and celebrity-friendly resort town of Castiglioncello.[1] Those whose interests veer more toward the gourmet than the literary and cinematic might also recall that one of the best red wines in the world (Sassicaia) is produced in this area, benefitting from the mild climate and the sea breezes.

Although far from being exhaustive, these kinds of cultural associations and representations of place have helped boost the appeal of this geographic area. Perceived by the majority of Italians and foreigners alike as an Arcadian, coveted tourist destination, the Etruscan Coast is usually appreciated for its well-tended beaches, unpolluted sea, natural parks, archeological sites, gastronomical excellence, and last but not least, progressive environmental policies.[2]

It goes without saying that this sort of Edenic, constructed perception is also substantially incomplete and does not capture the real and more complex nature of this place.[3] For example, on any given day, the air quality in Livorno's industrial harbor area is quite mephitic.[4] Tar, oil from ships, and plastic debris ranging from large bottles to microscopic spheres collect in even the most secluded rocky coves and beaches. Vada's famous "white beaches," with their Caribbean-like powder sand and eerie-looking turquoise waters, are actually the direct by-product of industrial toxic waste, loaded as they are with chemical compounds of ammonia, soda ash, and mercury from the local site of the Belgian multinational corporation Solvay, after which the "new town" of Rosignano Solvay was named.[5] However, at a time when Italy as a whole is facing skyrocketing unemployment rates and much larger economic and political problems, the reality of this environmental degradation, even on the rare occasions when it manages to reach a wider public, is generally considered an annoying yet mostly marginal curiosity.[6] This said, a poignant look at and realistic representation of the southernmost point of this coastal area, Piombino and its surroundings, can be found in a fictional contemporary literary work: Silvia Avallone's acclaimed first novel, *Swimming to Elba* (2012).[7] This novel problematizes the initial, idyllic picture I painted previously and simultaneously raises both local and global environmental concerns.

Set in 2001, *Swimming to Elba*'s main plotline tells the story of the struggling friendship, complicated existence, dreams, and coming of age of Anna and Francesca, two teenage girls who live in a working-class neighborhood in the industrial section of Piombino. Most of their relatives and friends work in the nearby Lucchini steel plant (once owned by the ILVA steelworks company). At the same time, the novel also deals with the close relationship and reciprocal interactions between human beings and nonhuman matter. This matter may be inorganic, like the iron ore and the machines the workers use to produce steel and the combination of concrete, asbestos, and rust with which their tenement houses are built, or organic, like the animals, plants, algae, and shells the girls find on their secret beach near the plant. These interactions are both literal and symbolic. For instance, human beings may assume thing-like qualities, like Francesca's father, who is significantly defined as a "whatsisname" (coso) after losing his finger, while objects may display uncanny anthropomorphic and vital qualities, like the fused metal that is subtly assimilated into the flow of blood or the blast furnace "Afo 4" that is described as a "vast burgeoning organism" (Avallone 2012, 16; smisurato organismo), animated twenty-four hours a day, which "digests, ruminates, and belches out" (16; digerisce, rimescola, erutta).

This section suggests that Avallone's novel not only presents an alternative vision of this part of Tuscany's famed Etruscan Coast but also illustrates the melding of substances through bodies and, in turn, the vulnerable imbrication of humans with their environs (and vice versa). By drawing attention to the mutual connections among organisms, ecosystems, and objects and among the biological, social, and technological spheres, this text goes far beyond the main story of Anna, Francesca, the industrial area of Piombino, and the Lucchini factory. Taking this fictional but also very real Tuscan territory as an example, Avallone's novel provides a template with which to better understand the intimate connections and the parallel degradation of human and nonhuman landscapes in many other coastal, rural, or urban "Piombinos"—be they nearby in Tuscany (the Solvay-owned soda ash plant in Rosignano comes again to mind), elsewhere in Italy, or anywhere on the globe.

My title, "A Life of Metal," hints at the homonymous chapter in Jane Bennett's *Vibrant Matter: A Political Ecology of Things*. The major claim of her book, which expands on a materialist tradition that follows the lines of "Democritus-Epicurus-Spinoza-Diderot-Deleuze more than Hegel-Marx-Adorno" (2010, xiii), is that there is an active vitality intrinsic to matter, that to separate what is "inert" from what is "vital" is more difficult than one may think, and that the agencies of nonhuman materials and forces ("operating in nature, in the human body, and in human artifacts" [xvi]) need to be considered to ultimately "counter the narcissism of humans in charge of the world" (xvi). The section of

Bennett's book titled "A Life of Metal" questions a traditional "association of metal with passivity or a dead thingness" and observes that "it is metal that best reveals this quivering effervescence; it is metal, bursting with a life, that gives rise to 'the prodigious idea of Nonorganic Life'" (55). As the earlier quotes from the novel suggest, Bennett's observation resonates well with some of the dynamics and situations described in *Swimming to Elba*.

Bennett's positions on material, distributed agency and her ideas about a posthuman connectedness of human and nonhuman entities emerging at both the corporeal and the discursive levels are shared by a number of other scholars who view the world as a place where everything is intermeshed and interferes with everything else, and where, by extension, the primacy and centrality traditionally assigned to human beings and their discourses at the expense of other forms of life and things (i.e., the anthropologocentric perspective) is radically questioned. Thus as a reminder, when Andrew Pickering states that we all live in the "mangle of practice" and that "everyday life has this character of coping with material agency, agency that comes at us from outside the human realm and that cannot be reduced to anything within that realm" (1995, 6), he wishes to surpass the dichotomies of society/nature, nature/culture, and human/nonhuman. Similarly, and despite her different disciplinary angle and emphasis, when Alaimo speaks of "trans-corporeality" to indicate "the extent to which the corporeal substance of the human is ultimately inseparable from 'the environment' . . . the flow of substances . . . between people, places and economic/political systems" and views the self as "a process of interacting agencies rather than a fixed, immobile and self-referential identity" (2010, 2–9), she shares an interest in moving beyond traditional juxtapositions of matter, agency, and meaning.

I suggest that *Swimming to Elba*'s relevance from an ecomaterialist perspective rests on its representation of various transformations of human and nonhuman bodies (be they those of the girls, the animals, the industrial products, or the land itself), on the interactions between these bodies and their environment, and on the mutual transformation of narrative, landscape, and nonhuman subjects. While depicting an intimate connection among organisms, ecosystems, and human-made substances, the novel also dramatizes the risky and potentially hazardous consequences that such a connection entails. That is, it illustrates the difficulty (if not the impossibility) of isolating the health or illness of geographic spaces from the health or illness of existential spaces. As the narrator asks,

> What does it mean to grow up in a complex of four big tenements shedding sections of balcony parts and chunks of asbestos into a courtyard where little kids play alongside older kids dealing drugs and old people who reek of decay? (Avallone 2012, 24)

> Cosa significa crescere in un complesso di quattro casermoni, da cui piovono pezzi di balcone e di amianto, in un cortile dove i bambini giocano accanto a ragazzi che spacciano e vecchie che puzzano?

Right from the start, Avallone's novel hints at this intersection between the biological and the social, at the mingling of bodies and world, and at the permeability of borders. The narration opens with the image of Francesca's attractive and sexually maturing body observed by her dull father (Enrico) from a distance as she walks along one of Piombino's beaches. Despite its fatherly provenance, there is no question that this is still a quintessential male gaze that, following an ancient and well-known trope, dissects and fragments Francesca's figure in detail, unavoidably objectifying her:

> Within the round blur of the lens, the body, headless, shifted slightly. A backlit wedge of flesh pulled into focus . . . muscles flexing just above the knee, the arc of the calf . . . a gleaming blond head of hair. . . . And the dimples on the cheeks, and the hollow between the shoulder blades, and the indentation of the belly button, and all the rest. (2012, 3)

> Nel cerchio sfocato della lente la figura si muoveva appena, senza testa. Uno spicchio di pelle zoomata in controluce . . . i muscoli tesi sopra il ginocchio, la curva del polpaccio . . . una splendida chioma bionda. . . . E le fossette sulle guance, e la fossa tra le scapole, e quella dell'ombelico, e tutto il resto.

In this passage, it is hard to miss the actual thingness of Francesca, her being introduced as belonging to a transitional ontological status between whole and part, human and thing ("headless"). Shifting attention to the beach where she is frolicking, this is a place where—as expressions such as "an ankle dusted with sand" (3; la caviglia sporca di sabbia), "[she] plunged into the water" (3; si iniettava dentro un'onda), and "the expanse of flesh pebbled with salt" (3–4; pelle intarsiata di sale) indicate—the sand, salt, and water stick to and mingle with her. But this is also where, at the same time and in a parallel dynamic, "the sand was mixed with rust and garbage; sewer pipes ran down the middle; and no one went there but criminals and the poor folk from the public housing" (4; la sabbia si mescolava alla ruggine e alle immondizie, in mezzo ci passavano gli scarichi, e ci andavano soltanto i delinquenti e i poveri cristi delle case popolari). Despite the fact that access to this beach seems ironically limited by one's (low) socioeconomic and (affirmative) penal status, all the substances at play cannot but intermingle, disregarding any real or constructed differences and hierarchies—any distinction between high and low. Thus the sand becomes a part of Francesca's attractive young body and mixes with the rust, the trash, and the industrial waste.

This melding of materials raises some questions: First, what may the direct or indirect consequences of this situation be? Second, who can tell with absolute certainty that the sand or any other matter coming from this degraded beach will not one day move around and, with its disturbing or potentially poisonous presence, contaminate the nearby shimmering "white beaches on the Isle of Elba . . . an unattainable paradise" (Avallone 2012, 10; spiagge bianche dell'isola d'Elba . . . un paradiso impossibile)? Finally, who or what will be eventually "swimming to Elba"? The novel eventually provides answers to some of these queries, but from these initial pages, it is already possible to suggest that the scenario Avallone is depicting is one where there cannot be pure, clear-cut distinctions and solid boundaries between different bodies or between inside and outside. In such a scenario, substances—be they organic or inorganic, natural or artificial—mix, and human beings are always entangled with both other subjects and other objects.

The narrative provides another literal instance of the mutuality between the human and the nonhuman by reminding readers that the large size and growth of Enrico's own body has been determined by external circumstances and materials: "From childhood he'd sculpted his muscles by hoeing and digging the earth. He had become a giant in the tomato fields, and, later, shoveling coking coal" (Avallone 2012, 5; Fin da bambino si era scolpito i muscoli a forza di zappare la terra. Si era fatto un gigante nei campi di pomodori, e poi spalando carbon coke). As it evokes the transition from agriculture to industry in Italy, this brief quote also establishes a parallel between human corporeality and that of the earth. On the one hand, it reminds us that the development of the former is inescapably and organically tied to the agentic substances produced by the latter. On the other, especially if considering our previous observations on the "fatherly gaze," it speaks of the potential risk that arises whenever there is a gigantic, anomalous, "ogre-like" overgrowth of a (male) human being present.

The following chapter of *Swimming to Elba* takes us inside the Lucchini factory, where one finds even clearer examples of corporeal and material entanglements. The opening scene is of a fistfight between Alessio (Anna's brother) and another younger factory worker who disrespected Anna. Here Avallone informs us not only that this factory cultivates and projects violence, energy, and power but also that it is a place where distinctions between human and nonhuman, inert and living matter, are particularly hard to make. Thus in the same way that inorganic substances, like the iron filings Alessio swallows during his fight ("He had to swallow a huge gob of spit and iron filings to keep calm" [Avallone 2012, 14–15; Dovette ingoiare un bolo grosso così di saliva e limatura di ferro, per restare calmo]), define and become an integral part of the human subject, organic matter is processed by the body of the plant and ends up taking part in the creation of its final product.

Bennett's expression of "material vitality" (or "vital materiality" [2010, 60]), indicating "the elusive idea of a materiality that is itself heterogeneous, itself a differential of intensities, itself a life" (57), is especially fitting in the context of the industrial factory and its various theriomorphic components, unfailingly described by Avallone as being a "burgeoning" (2012, 16) animated organism. A word like *secretion*, more commonly associated with the physiological sphere, is used to depict the multiple hybrid agencies involved in the production and "birth" of steel:

> Steel does not exist in nature, it is not an elementary substance. It's a secretion of thousands of human hands, electric meters, mechanical arms and every so often the skin of a cat that's tumbled into the molten alloy. (4)

> L'acciaio non esiste in natura, non è una sostanza elementare. La secrezione di migliaia di braccia umane, contatori elettrici, bracci meccanici, e a volte la pelliccia di un gatto che ci finisce dentro.

Metal, continues Bennett, "is always metallurgical, always an alloy of the endeavors of many bodies, always something worked on by geological, biological, and often human agencies" (2010, 60). Multiple agencies and the secretions of many bodies are certainly involved here—to a point where human and mechanical arms, electricity, and even some cat fur become the necessary ingredients to be mixed and swallowed by the ladles where the melting takes place. Such an enormous container is just the first surreal inhabitant of this fantastic, transcorporeal ecosystem. As Avallone explains,

> It's an entire zoo in here: pig iron everywhere, cranes of every species and variety. Rusted animals with horned heads. . . . The dense black sludge of molten metal was bubbling in the crucibles, potbellied swiveling barrels running along on the mill trains. Giant tanks on wheels that looked like primordial creatures. . . . Metal was everywhere, in the process of birth. Unceasing cascades of steel and glistening cast iron and viscous light. Torrents, rapids, estuaries of molten metal coursing through the flow lines, into the ampules of the ladles and pouring out into the tundishes to drain into the molds for furnaces and trains. . . . Raw materials were being transformed at every hour of the day and night. . . . You could feel the blood rushing through your arteries at a fantastic velocity in there, and from the arteries to the capillaries, while your muscles built up in tiny fractures: You were regressing to the animal state. Alessio was small and alive in this vast burgeoning organism. (2012, 15–16)

> Un intero zoo: nel cielo svettavano torri merlate, gru di ogni genere e specie. Animali arrugginiti dalle teste cornute. . . . La melma densa e nera del metallo fuso ribolliva nelle siviere, barili panciuti trasportati dai treni siluro. Cisterne munite di ruote che assomigliavano a creature primordiali. . . . Il metallo era ovunque, allo stato nascente. Ininterrotte cascate di acciaio e ghisa lucente e luce vischiosa. Torrenti, rapide, estuari di metallo fuso lungo gli argini delle colate e nelle ampolle dei barili. . . . A ogni ora del giorno e della notte la materia veniva trasformata. . . . Ti sentivi il sangue circolare a ritmo pazzesco, là in mezzo, dalle arterie ai capillari, e i muscoli aumentare in piccole fratture: retrocedevi allo stato animale. Alessio era piccolo e vivo in questo smisurato organismo.

The various references to animals, streams, rapids, and estuaries on the one hand and human arteries, capillaries, and muscles on the other reveal Avallone's wish to rhetorically correlate different bodies that are traditionally considered separate and to analogically assimilate the artificial landscape of technology and the processes of industrial production to natural and physiological phenomena. In fact, the awe-inspiring energy, excitement, and dangerous beauty that characterize this place strongly evoke the aesthetic notion of the "technological sublime" in a peculiar postmodern (and posthuman) hybrid manifestation.[8] In other words, this is a manifestation where nature—or, better yet, the language of nature—has not been completely displaced by the machine but is still reappearing and is exploited to convey a striking experience. At the same time, Timothy Morton's reminder that "nature as such appears when we lose it, and it's known as a loss" (2010, 133) makes us realize that this resurfacing and discursive presence of nature is nothing but a spectral trace, a linguistic remnant and, as such, also the sign of an actual loss, an absence. As a consequence, one cannot be sure of the location of borders or the meaning of nature; tangentially, readers learn that "it is impossible to separate matters of social and environmental concern from discursive ones" (Iovino and Oppermann 2012, 463).[9]

"Where does the body end and 'nonhuman nature' begin?" Alaimo asks in her *Bodily Natures* (2010, 11). *Swimming to Elba* provides a potential answer by implying that technology, animality, landscape, and humanity are all intermeshed. This particular human being, Alessio, like every other thing, is just a part of this immense, more-than-human organism. His substance is "inseparable from the environment" (Alaimo 2010, 2), and his life and vitality are dependent on (and even enhanced by) the life and vitality of the steel plant that, in turn, depends on him to function properly.[10] Various instances of the vibrancy of matter and its transformations and of its multiple levels of agency thus surface throughout Avallone's work:

> [Alessio] shot a quick glance at the blonde on the Maxim calendar. Constant yearning for sex, in the mill. The reaction of the human body inside the titanic body of industry: It's not a factory but material changing form. (2012, 16)

> Diede un'occhiata alla bionda del calendario Maxim. Perenne desiderio di scopare, là dentro. La reazione del corpo umano nel corpo titanico dell'industria: che non è una fabbrica, ma la materia che cambia forma.

If taken literally, these lines suggest not only that a factory and a human being are both bodies constituted by and manifesting different shapes and degrees of material animation but also that it is through their respective agencies—their mutual intercourse—that a final product (be it industrial/inorganic, like steel, or biological/organic, like an actual secretion) may be completed.[11] Significantly, although Alessio is aroused by looking at the image of a pinup blonde on a calendar, the text suggests that the monstrous nonhuman body of the plant blast furnace is the real, though involuntary, target of his physiological reaction, since he is, after all, responsible for "inseminating" it so that it can deliver its final product: "The artificial insemination took place in a test tube as tall as a skyscraper, the rust-flaked urn of Afo 4 that has hundreds of arms and bellies and three horns in place of a head" (16; La fecondazione assistita avveniva in un'ampolla alta come un grattacielo, l'urna rugginosa di Afo 4 che ha centinaia di braccia e pance, e un tricorno al posto della testa).

In other words, the textual closeness between Alessio's sexual yearning and the action of artificial insemination that sparks the creation of steel can hardly be random. Not surprisingly, Afo 4's animallike features and physiological behaviors along with Alessio's hybrid corporeality are further underlined in the following paragraphs, together with a clear hint at the hazardous by-products and catastrophic effects created by such an erotic human-machine interaction:

> He could feel the pressure of it at the nape of his neck, the black tower of the Afo 4, a giant spider that digests, ruminates, and belches out. . . . Bluish fluorescences, toxic clouds in volumes sufficient to poison not only the Val di Cornia, but all of Tuscany. . . . Tons of metal whirling like birds, yellow clouds of carbon smoke, black at the mouths of the smokestacks. It's called continuous integrated steel production. As Alessio walked he crushed nettles and chunks of refractory brick underfoot. Metal saturated the ground and his skin. . . . The elementary motion of machinery that is no different from life. (Avallone 2012, 16–17)

> Se la sentiva premere sulla nuca, la torre nera di Afo 4, il gigantesco ragno che digerisce, rimescola erutta. . . . Fluorescenze azzurognole, nubi tossiche in quantità sufficiente ad ammorbare non solo la Val di Cornia, ma la Toscana intera. . . . Tonnellate di metallo vorticavano come uccelli, nuvole gialle di carbonio, nere dalle bocche delle ciminiere. Si chiama ciclo continuo integrale. Alessio calpestava ortiche e mattoni refrattari. Il metallo saturava il terreno e la sua pelle. . . . Il movimento elementare della macchina che è uguale alla vita.

What captures one's attention in this literal and metaphorical scenario of contamination is the intimacy of Alessio's and Afo 4's different and yet physically proximate bodies, the indication of the various and ubiquitous embodiments of metal (and other materials), and the explicit analogy and direct attribution of life to the technological object. At another level, these lines evoke a monstrous, toxic birth[12]—what Rosi Braidotti has "coded as *zoe* . . . the nonhuman, vital [generative] force of Life . . . the transversal force that cuts across and reconnects previously segregated species, categories and domains" (2013, 60), which implies an expanded, postanthropocentric notion of the relational self and also alludes to Thanatos, *zoe*'s complementary "other side," the "death-bound or necro-political face of post-anthropocentrism" (118). Braidotti writes, "Life as *zoe* also encompasses what we call 'death'" (134). The objective of her "posthuman affirmative ethics" is to stress the positive, "productive aspect of the life-death continuum" and "the politics of life itself as a relentlessly generative force including and going beyond death" (121). Namely, from her posthuman, vitalist brand of materialism, "the emphasis on the impersonality of life is echoed by an analogous reflection on death," and the latter "could not be further removed from the notion of death as the inanimate and indifferent state of matter, the entropic state to which the body is supposed to 'return'" (131–37). Thus death, like life, "is not a human prerogative" (137). However, at least in the context of Avallone's novel, it is hard to individuate any of death's "productive aspect[s]," and one undoubtedly sees more of the negative, lethal effects of industrialized modernity on multiple subjects and objects.

Indeed, *Swimming to Elba* repeatedly evokes a situation of extended, transversal vulnerability. Alessio's encounters with various organic and inorganic diseased bodies—"the rotting corpse of a rat" (Avallone 2012, 16; il cadavere putrefatto di un topo), "the posthumous carcasses of the three blast furnaces that hadn't yet been dismantled" (17; le carcasse postume dei tre altoforni non ancora smantellati), and the local population of resident cats, "all of them mangy and sick, all of them calico, with black and white patches, from the relentless inbreeding" (18; tutti malati, tutti bianchi e neri a forza di incrociarsi sempre tra loro)—suggest that everyone and everything is eventually perishable, regardless of its ontological status. Just like life, death, destruction, and illness are not the prerogative of just one thing. As Braidotti observes, "The body [any body] doubles up as the potential corpse it has always been" (2013, 119).

In the advanced capitalist model that the surviving Lucchini plant embodies, one where, Avallone notes, "the West was reproducing the world and exporting it" (2012, 18; l'Occidente riproduce il mondo e lo esporta), vitality and decay, beginning and end coexist and affect multiple corporealities. The apocalyptic landscape described in the novel perfectly captures the complex dynamics and encompassing proximity between inorganic death and organic life and vice versa:

> Some sectors of the plant were dying; smokestacks and industrial sheds were being demolished with dynamite.... Still they, the seventh-generation millworkers, had their fun, riding the power shovels like bucking bulls, their transistor radios blaring out at full volume, an amphetamine tablet dissolving under their tongues. (18)
>
> Alcuni rami della fabbrica morivano, ciminiere e capannoni venivano fatti saltare con il tritolo ... gli operai ... si divertivano a cavalcare gli escavatori come tori, con le radioline portatili a palla e una pasticca di anfetamina sciolta sotto la lingua.

Alessio's own absurd death at the end of the novel reminds us yet again of the entanglements and environmental interconnections addressed so far. In fact, similar to a dismantled blast furnace, he represents a human "posthumous carcass" (17; carcassa postuma; recalling one of Braidotti's "potential corpse[s]")—another disposable part of the plant. Yet as his own corporeality vanishes, he becomes indistinguishable from and merged with the posthuman body of the plant: "Alessio had ... stopped being a human body ... had become a puddle of blood, spreading between the scattered steel rod, a blinding pool. Not Alessio. A cat" (295; Alessio aveva ... cessato di essere un corpo, ed era diventato ... una pozza di sangue allargata tra i tondi, una polla abbacinante. Non Alessio. Un gatto).

The novel's initial chapters are not the only ones that benefit from a material ecocritical analysis. References to the dangerously attractive vibrancy of matter and to the corporeal interchanges among "human bodies, nonhuman creatures, ecological systems, chemical agents, and other actors" (Alaimo 2010, 2) recur in significant moments in the narration. When, for example, Anna and Francesca hop on a scooter and ride to an abandoned area within the Lucchini plant—another "dead branch of the factory [that] had been reduced to a carcass of rusted metal ... [an] industrial cemetery" (Avallone 2012, 53; ramo morto della fabbrica [che] si era ridotto a una carcassa di ruggine ... [un] cimitero industriale)—the juxtaposition between the burning sexual vitality of the adolescents and the lifeless, scorching condition of the place is hard to miss. However, one soon realizes that the terms of such a juxtaposition can be reversed, since multiple substances coming from the plant are entering and vitally affecting everyone's bodies, turning the friends into endangered appendages of this hellish ecosystem:

> You could feel the light coating of coal dust filtering into your lungs, sticking to your body, blackening your skin.... The body was pounding hard along with the metals in

the furnaces. Rebar, slabs, billets: along with the heart, the arteries, the aorta. . . . The incessant, raucous lament of the steel mills—you could feel it vibrating deep in your bones. . . . Lead and the heavy scent of iron burned the lungs and the nostrils . . . you could feel yourself liquefying. (52–53)

Lo spolverino prodotto dal carbone te lo sentivi entrare nei polmoni appiccicarsi addosso, annerire la pelle. . . . Il corpo batteva forte insieme ai metalli nei forni. Le barre, i blumi, le billette: insieme al cuore, le arterie, l'aorta. . . . Il lamento rauco, perenne delle acciaierie, te lo sentivi vibrare nelle ossa. . . . Il piombo, l'odore pesante del ferro bruciava i polmoni e le narici . . . ti sentivi liquefare.

Once again, human and nonhuman matter intermesh until substance-saturated human bodies, just like the metals involved in the production of steel, may liquefy. At that time, words like *dead* and *burning* might describe both a few pieces of anthropomorphized machinery and, from a posthumous perspective, also those young human bodies. Enrico's own body, with his fractured ribs and vertebrae and a shattered hand, is not only broken (just like the excavator) because of his job-related accident but literally transformed into a useless thing ("Then he wasn't even a first and last name anymore. *Whatsisname*, he'd become" [Avallone 2012, 248; Poi, non era più neanche un nome e cognome. *Coso*, era diventato]). Moreover, his amputated finger mirrors a physical loss in the industrial body of the plant:

The decline in the price of steel on the world market over the past two decades had forced the steelworks to dismantle Afo 1, Afo 2, and Afo 3. They were gone now. Like his finger. An enormous hole in the toxin-saturated earth. (248)

La flessione dell'acciaio sul mercato, nel giro di due decenni, aveva costretto a smantellare Afo 1, 2, e 3. Non c'erano più. Come il suo dito. Un buco enorme nel terreno saturo di veleno.

As Enrico's finger vanishes, anticipating by synecdoche the disappearance of both the blast furnaces and Alessio, he becomes "a zero in the depressed system" (249; uno zero nel sistema depresso)—another discardable entity in this inhuman, market-driven space of corporeal confusion and contamination.

An apparent natural opposite of the factory is the girls' secret beach ("the canebrake" [Avallone 2012, 94; il canneto]). I say *apparent* because, despite the obvious differences, there is also a certain similarity between these two crucial locations. They are sites where human and nonhuman elements coalesce, where equal importance is assigned to both subjects and objects, and where the vitality and mortality of matter are made evident. Animate and inanimate things, waste products, and debris and garbage are all constitutive parts of this ecosystem and convey an essential continuity and mutuality between the industrial/technological and the supposedly natural. Thus in the canebrake, Francesca and Anna, "emitting little, scarcely human bursts of giggles" (94; emettendo risolini poco umani), become just "a pair of excrescences on that murky landscape" (94; due escrescenze di quel luogo), their sweat mixing with the sap from the vegetation, their soiled bodies merging with those of the insects, algae, shells, cats, and all the other things, either alive or dead, that end up here:

> Their damp hair, scented with shampoo, was gradually being impregnated with another odor: a mixture of sweat and sap. The downy growth on the surface of the plants caused their skin to itch. It felt as if they were walking through wool. . . . The beach was piled high with seaweed. . . . Fishermen came here to leave the broken hulks of boats to save money on dumping fees. . . . Balls of neptune grass by the million, all tossed up by the sea right there. On the shore they broke down into a black mucilage, a muck that reeked of pee and fresh bread. . . . They chewed the algae. They sank their faces into the damp and rough furs of the cats. . . . That dead zone of the coastline was reduced to a primordial broth of things . . . a ladle, a ceramic tile. [Francesca]'d hunker over her digging and shout if she unearthed anything human. (94–95, 96)

> Sudore misto a linfa. La peluria delle piante pruriginava a contatto con la pelle. Sembrava di camminare nella lana. . . . La spiaggia era un cumulo di alghe. . . . I pescatori ci venivano a gettare le carcasse per non pagare la tassa dei rifiuti. . . . Posidonie brune a milioni, riversate dal mare tutte lì. Sulla riva si sfibravano in una mucillagine nera, una poltiglia che sapeva di pipì e di pane. . . . Masticavano le alghe. Affondavano il muso nelle pellicce umide e ruvide dei gatti. . . . Quel punto morto della costa era ridotto a un brodo primordiale di cose . . . un mestolo, una piastrella di ceramica. [Francesca] Si chinava a scavare e gridava, se dissotterrava qualcosa di umano.

The ironic contrast emerging in these lines between the lifeless condition of this abandoned, dead coastal area and the fact that, at the same time, it also constitutes a "primordial broth" cannot be overlooked. Not only is this apparent necroregion[13] at odds with its definition as the hot, carbon-based solution that is allegedly at the origin of organic life on earth, but such a life-giving substance is paradoxically constituted by human-made things and inorganic debris. In other words, these are objects that simultaneously suggest their condition as both dead and alive, or as Bennett puts it, they "at one moment disclos[e] themselves as dead stuff and at the next as live presence: junk, then claimant; inert matter, then live wire" (2010, 5).[14]

This condition reemerges in a later, related passage. After her falling out with Francesca, Anna revisits the beach alone and reminisces while gazing at the garbage deposited by the sea. Avallone describes the scene:

> Empty gas canisters, used sanitary napkins, glass and plastic bottles. . . . She walked by the cadavers of things. There were broken dishes and fruit juice cartons. There were plastic spoons and forks and shredded plastic plates. The rusty shower up there, and here a broken toy pail. (2012, 221)

> Cisterne vuote, assorbenti usati, bottiglie di plastica e di vetro. . . . Passava accanto ai cadaveri delle cose. C'erano cocci e cartoni di succo di frutta. C'erano posate e piatti di plastica sventrati. Le docce arrugginite lassù, e qui un secchiello rotto.

This sight deeply affects Anna. This vision of desolation and abandon stirs her memory, increases her awareness of the effects this place has on her: "Places cling to you. Places become extraneous to you" (222; I luoghi ti impastano. I luoghi ti diventano estranei). Most significantly, it makes her realize that there are "the things that come back" and "the things that can never come back" (222; le cose che ritornano e le cose che non possono tornare), implicitly associating Francesca (and their interrupted friendship) with one of those moving things. Once again, therefore, Bennett's observation that "a vital materiality can never really be thrown 'away,' for it continues its activities even as discarded or unwanted" and her affirmation that inanimate things have the "curious ability . . . to animate, to act, to produce effects dramatic and subtle" (2010, 6) aptly describe the kind of interactions between human and nonhuman entities mentioned earlier.

At this point, it is superfluous to stress how Avallone's novel understands the life and mind of this particular place—how her work is informed by a marked sensibility to and awareness of the environmental, socioeconomic, and political issues that characterize Piombino's industrial territory and its surroundings. Nor should I need to underline the extent to which the narration alludes to the kind of coextensiveness across bodies and blurring among people, things, and landscapes that have been theorized by scholars of vital materialism and posthumanism. Therefore, I prefer to conclude by reflecting on a few of the larger ecological implications of this closeness and mutual relationship between organic and inorganic substances.

First, while *Swimming to Elba* depicts the specific dynamics of this local Tuscan landscape, it also evokes comparable global capitalist practices of industrial production and consumption and shows their real effects on other human and nonhuman bodies. Second, while doing so, Avallone reminds us of literature's power to provide readers with knowledge, awareness, and "the critical instruments necessary to develop their own 'strategy of survival' both environmentally and politically" (Iovino 2009, 343). Serenella Iovino writes that a story may initiate the practice of "restoring the imagination of place" only when it is "open"—that is, when it shows "awareness (about values and critical issues), projectuality (vision of the future), and empathy (as a mutually enhancing dialectic amid different subjects)" (Iovino 2012b, 107). Even though ecological questions may not be its most obvious or principal subject, *Swimming to Elba* is "open" in the way Iovino describes and, as such, is an example of what she calls "narrative reinhabitation." This enables the work to affect a taken-for-granted knowledge of the world and reveals that it has the potential to transform a (quasi) necroregion into an "evolutionary landscape" (112).

In its final pages, the novel hints at what such an evolution may look like in Piombino's industrial area by alluding to some of the changes the Lucchini plant will likely undergo:

> People were already starting to talk about reclamation, about dismantling the steel mills. Converting the local economy, focusing on tourism, the service industry. . . . Like the Colosseum, like the hulls of the fishing boats stranded on the sand, the blast furnace, too, in a decade or so, would become the property of the cats. (Avallone 2012, 303)

> Cominciavano già a parlare di bonifica, di smantellamento. Convertire l'economia locale, puntare al turismo e al terziario. . . . Come il Colosseo, come gli scafi arenati sulla spiaggia, anche l'altoforno, nel giro di un decennio, se lo sarebbero preso i gatti.

Readers learn that the substances that constitute this place and those that make the humans who inhabit it are not that different: what happens to one may eventually happen to the other. It should not be surprising, therefore, that in the end, Francesca—now described as "the most radiant of all the elements" (308; il più radioso fra gli elementi)—travels with her best friend, Anna, from Piombino to Elba, symbolically taking the place and reversing the trip of the iron ore that was once shipped from Elba's mines to Piombino's industries.[15] While this action can be seen as a first step in the right direction toward such a posthuman form of restoration and reinhabitation, it cannot but recall and further unsettle the idyllic view of the Etruscan Coast I initially sketched. As the text reimagines and redraws the borders between these Arcadian and industrial spaces, it reminds us that any landscape requires more complex and hybridized readings than those we are usually exposed to. The "blue flag" I mentioned at the beginning may now well show a hint of steel gray.

Seeking Environmental Justice in Italian Asbestos Narratives

> There is no steel without asbestos. . . . I was born between Casale's asbestos and Piombino's steel, between killer dusts and the castings of blast furnaces.
>
> Non esiste acciaio senza amianto. . . . Sono nato tra l'amianto di Casale e l'acciaio di Piombino, tra la polvere assassina e le colate degli altiforni.
>
> —Alberto Prunetti, *Amianto: Una storia operaia*

> Every mine is magic. . . . Earth's depths teem with elves . . . who can either be generous and let one find a treasure under the tip of the pickaxe or deceive and bedazzle.
>
> Già tutte le miniere sono magiche. . . . Le viscere della terra brulicano di gnomi . . . che possono essere generosi e farti trovare il tesoro sotto la punta del piccone, o ingannarti, abbagliarti.
>
> —Primo Levi, *The Periodic Table*

While discussing the ILVA-owned steel plant in the city of Taranto in the southern region of Puglia, Monica Seger points out how this industrial-sized iron metallurgy regularly releases huge amounts of highly toxic dioxins and dioxin-like compounds that pollute the environment and, over time, endanger one's health (2018, 185).[16] If dioxin is among the natural by-products of the steel-making process, one might argue that asbestos—also known as the "magic mineral"—has long been necessary to make it happen. Until 1992, when asbestos's production and usage were banned in Italy, it was ubiquitously utilized in steel plants and many other industries (automotive, construction, electrical, plumbing) as a crucial ingredient of compounds and as an effective insulation material. Already known in ancient times, its name derives from the Greek *amíantos* (Latin, *amiantum*), meaning "incorruptible," to indicate its very high resistance to fusion and combustion. At the same time, it means that it is eternal— or as Nicky Gregson, Helen Watkins, and Melania Calestani put it, it does not "go away once declared to be hazardous. . . . [It has a] future" (2010, 1067).

Classified as a fibrous mineral, it is found in nature on almost every continent of the globe, and it is usually extracted in open-pit mines, liberated from the surrounding rocks through mechanical and chemical procedures of grinding and enrichment.[17] As early as the 1970s, scientist writer Primo Levi, author of *The Periodic Table*, perceived the double nature of this substance— depending on circumstances, either benign or malignant—while providing a vivid and detailed picture of one of its original sites of production in Northern Italy. Because Levi's story is set in 1941, asbestos's hazardous and carcinogenic toxicity is not mentioned in his description of the Balangero mine in Piedmont, where he worked as a young chemist testing new extraction techniques. Yet what he tells us about this place—namely, that it is the largest mine of its kind in Europe and reminds him of Dante's hell in his *Divine Comedy*, with its conic circles and a "monstrous mill" (mostruoso frantoio) that is as noisy as "the apocalypse" (1984, 68; un fracasso da apocalissi)—helps us form a preliminary idea of both this substance's propensity to be everywhere and the various kinds of expenditures—human, material, and economic—that this entails. As Levi writes,

> The ultimate goal of that cyclopean work was to wrestle a sparse 2% of asbestos from the rocks that trapped it inside. . . . Asbestos was everywhere, like ashen snow: if one left a book on a table for a few hours, and then lifted it up, its negative profile was outlined; the roofs were covered with a thick layer of dust. (65)
>
> Scopo ultimo di quel lavoro da ciclopi era strappare alla roccia un misero 2 per cento d'amianto che vi era intrappolato. . . . C'era amianto dappertutto, come una neve cenerina: se si lasciava per qualche ora un libro su di un tavolo, e poi lo si toglieva, se ne trovava il profilo in negativo; i tetti erano coperti da uno spesso strato di polverino.

> Our fascination with buried wealth, with two kilos of noble, silver metal, linked to one thousand kilos of sterile stone which is thrown away has not yet ended. (78)
>
> Il fascino della ricchezza sepolta, dei due chili di nobile metallo argenteo legati ai mille chili di sasso sterile che si getta via, non si è ancora estinto.

As a chemist and a self-described "hunter" (cacciatore) who is always searching for ways to "find an opening" (cercare il varco) into nature's complexity and the "incomprehensibility of matter" (79; materia incomprensibile), Levi ultimately remains more intrigued by the many techno-scientific challenges he faces at the Balangero mine than concerned about its socioenvironmental impact.

Such an impact is nonetheless much more central in a short unpublished essay by another famous Italian writer, Italo Calvino, who visited the same mine as a reporter for the daily newspaper *L'Unità* in 1954. In his article "La fabbrica nella montagna" (The factory in the mountain), as the workers and miners approach the factory from the surrounding mountain paths, "with their fustian jackets, boots, caps and balaclavas" (1954, 192; con le giacche di fustagno, gli scarponi, i berretti col passamontagna), they remind him of unarmed hunters. However, one soon learns that these men are entering a necrozone where "there are no hares in the woods, no mushrooms grow . . . no corn grows. . . . There is only the gray dust cloud of asbestos that as far as it gets burns leaves and lungs, there is the pit . . . their life and death" (192; non ce nè di lepri nel bosco, non crescono funghi . . . non cresce frumento. . . . C'è solo il grigio polverone d'asbesto della cava che dove arriva brucia, foglie e polmoni, c'è la cava . . . loro vita e loro morte). In effect, the reader realizes what the actual killer is and, in turn, who the victims in this place are, as agency is subtly shifted from human beings to the excavated material.[18]

Innocuous and useful when inert and undisturbed, asbestos becomes almost as dangerous as dioxin to any living creature that inhales its nano-sized, needlelike fibers when it is broken and fragmented. And like dioxin, asbestos has similarly delayed health effects on those exposed to it. The kind of cancer it induces (mesothelioma) and the disease of asbestosis are like slow-burning fires that may appear decades after initial contact.[19] One might reasonably speculate that Rob Nixon was also thinking of asbestos and its "delayed destruction" (2011, 2) capabilities when the notion of "slow violence" first came into his mind (see also Iovino 2016a, 152–53). As its material state changes from inert and stable to active because of either some human practice and interference or natural causes (e.g., decay, floods, fires), so do its properties and nature, the latter shifting over time from that of an almost benign, protective substance to a ticking time bomb—a "lethal serial killer who [in Italy] has been operating in Casale Monferrato, Taranto, Piombino and tens of other places" (Prunetti 2012, 113; serial killer micidiale che agiva a Casale Monferrato, a Taranto, a Piombino e in decine d' altri posti).

In an intriguing paper about the release of asbestos fibers during shipbreaking procedures, geographer Gregson and her coauthors explore from a vital materialist perspective asbestos's "storied materiality" (Gregson, Watkins, and Calestani 2010, 1068); its performative, ontological copresence alongside humans; and some of the implications of the transformation and animation mentioned earlier. The authors note that asbestos is often reanimated through the practice of demolition. In particular, by combining materials and human activity, demolition is defined as a "state of flux . . . a human-orchestrated re-meshing of materials in the world" that is accompanied by simultaneous "practices of salvage and remediation." Additionally, they remark that asbestos's performative vitality is likely to be best exemplified "in its co-presence within a human body with asbestosis, struggling to breathe, and the horrible, drawn out, lingering death of mesothelioma" (1081).

In combination with Levi's and Calvino's fictional pieces, Gregson and her colleagues' critical observations thus form an appropriate introduction to the discussion of more recent "narratives of matter" (or, alternatively, as Levi writes, "mineral tales" [1984, 78; storie minerali]). By addressing the embodied manifestations, entanglements, and toxic legacy of asbestos in different bodies, places, and textual sites, these modern narratives promise to complement from a discursive standpoint Gregson and her coauthors' empirical approach. Such narratives, like several others we have encountered so far, "loop between material world and textual reality" (Iovino 2012a, 62), narrowing the perceived gap between them. In the process, terms and notions such as *demolition, remediation, salvage,* and *recovery* acquire additional layers of meaning that, as cultural ecology theorizes, relate in different and yet intertwined ways to a physical dimension as well as to a more abstract, ideal, and sociocultural one. To put it otherwise, from this perspective, storytelling may be viewed as another "[state] of flux" that "re-mesh[es] materials in the world" (Gregson, Watkins, and Calestani 2010, 1080). The future of asbestos materializes posthumously both corporeally every time illnesses manifest themselves and discursively whenever the substance resurfaces in and animates the cultural, linguistic, and literary spheres. At the same time, writing becomes part of a creative process of cognitive reconstruction that—now imaginatively mixing asbestos's fibers and human creativity—counters demolition and attempts, as far as possible, to remediate, salvage, and recover. The act of narration becomes thus not only an action of cultural remediation and memory against forgetting but also a vehicle to make sense of and contain future risks scenarios, revitalize searches for environmental justice, and visualize and give corporeal weight to those invisible fibers. This will "make it more difficult for future generations to maintain that 'there was nothing ever said, nobody knew'" about asbestos exposure and impact (Lindgren and Phillips 2016, 159).

Both a "human body with asbestosis" and its eventual "death by mesothelioma" as well as the notion of writing as a healing process are at the core of Alberto Prunetti's *Amianto: Una storia operaia* (Asbestos: A blue-collar story; 2012), the first of these "narratives of matter" I wish to consider. This novel is part of an increasing number of literary and cinematic productions that have contributed to recognizing and memorializing thousands of victims of asbestos-related deaths while also increasing public sensibility toward this issue. A mixture of linguistic and material registers characterizes the text; in fact, the story is narrated "mixing documents and memories as if [they] were blended in 'lean concrete' [*magrone*], that coarse kind of mortar mixed with the gravel of my childhood" (Prunetti 2012, 131; mescolando ricordi e documenti come nel 'magrone,' quella malta grossolana impastata con la ghiaia della mia infanzia). As the narrator revealingly tells us, "For me memory is alive and may be fed in many ways [even] by passing a grinder on the rust of time" (112; La memoria per me è una cosa viva che si preserva passando la smerigliatrice sulla ruggine del tempo).

Written in a hybrid form at the crossroads of (auto)biography, environmental record, medical bulletin, labor report, and social justice narrative, the novel recounts the corporeal decay and eventual death of Renato, the narrator's father, a nomadic blue-collar worker who began his career in Piombino and later traveled from his native Tuscany throughout Italy to perform welding and pipefitting at other industrial sites.[20] In doing so, *Amianto* not only addresses intergenerational issues of trauma and remembrance but also reveals the toxic legacy of asbestos that (mostly) burdens the working class. As Monica Jansen points out, the novel also conveys the "tragic awareness that labor . . . has become . . . an occupational practice that debilitates and eventually extinguishes the working class as well as its single components" (2017, 6). Jansen's joint discussion of *Amianto* and Stefano Valenti's *La fabbrica del panico* (The panic factory; 2013)—another "asbestos novel" dealing with labor-related traumas (disease, death, precariousness)—draws on theories of affect and postmemory. Jansen argues that these narratives are not just factual, objective mirrors of the outside world and that, by adopting transmedial forms of storytelling, they attempt to impact reality by turning traumatic feelings such as anger, betrayal, and hurt "into cohesive sentiments, such as solidarity and sociability, which may foster redemption and change" (Di Martino and Verdicchio 2017, xvii). As Jansen summarizes toward the end of her essay, while private events in these narratives become collective and emblematize a "resistance against capitalism . . . the act of narration [is expanded] into that of cultural activism" (2017, 22–23).

In the following pages, my objective is to complement and corroborate Jansen's insightful remarks by insisting just a bit more, from a different angle, on the "recalcitrant materiality" (Breu 2014, ix) and semiotic power of the substance that gives the title to Prunetti's novel. In particular, I consider Prunetti's *Amianto* in light of Alaimo's observations on "material memoirs" (2010, 95) while also being mindful of some of Marco Armiero and Ilenia Iengo's (2017) remarks on the double aim (personal and social) of stories about illness. I also keep present Arne De Boever's elaborations of literature as a *pharmakon* and of "sickness narrative" as "a work of bioart that is traversed by both biopolitical concerns and by concerns with the care for the self and the care for others"—a genre complicit with "biopolitics," or "a politics *over* life" (2013, 13), but also able to "seize the possibilities *of* life-narrative as a technique of emancipation and liberation (one could call this a biopolitics *from below*—a politics of life)" (154). Prunetti's "life-narrative" would seem to encourage this sort of multipronged approach, since he tells of the transcorporeal coextension among asbestos, a blue-collar worker's body poisoned by it, and capitalism. Moreover, as he observes in the paratextual conversation with authors Wu Ming 1 and Girolamo De Michele at the end of the novel ("Il Triello"), his book "is a scar, a scar that was healed by the suture thread of writing" (2012, 161; una cicatrice, una cicatrice che si è rimarginata col filo di sutura della scrittura).

On the one hand, an invisible, immaterial asbestos—perhaps more than other substances—iconizes the time-bound hazards of industrial capitalism and, as such, is imbricated in the latter's bio- (and thanato-) political practices against more vulnerable bodies and social classes. Yet on the other, asbestos's visible textual rematerialization and manifestation in Renato's increasingly sick body is also a healing and potentially liberating gesture. Arguably, while the novel reanimates the memory of a dead subject, it also attempts to deanimate asbestos's vitality and defuse its hazardousness for everyone else who is still living. In this limited sense, writing functions as a sort of shield and "curative" practice that protects other Renatos from similar deadly exposures in the future. And continuing to play with asbestos's double nature—if as a material it may be viewed as a "technology" of capital (Sudan 2016, 7) and, as such, inherently complicit with capital's biopolitical threats—when associated with a sick body and viewed as a literary trope, it may actually help prevent or mitigate "integration with the world of biopolitical control" (Breu 2014, 2).

While discussing counterhegemonic narratives of illness in Campania, Armiero and Iengo observe that "ill bodies enact different form of resistance through storytelling" (2017, 55). As Prunetti retraces his father's and his own lives, going over a family history where "bodies and selves are constructed from the very stuff of the toxic places they have inhabited" (Alaimo 2010, 102), he transforms his personal encounter with asbestos-induced sickness into a shared, public, collective one. Through storytelling, his work participates in what Armiero and Iengo call the "politicization of ill bodies." This process, as the two scholars observe, "is inscribed in feminist practice that considers the personal to be political . . . [and] occurs when a personal sufferance becomes an experience revealing the patterns of injustice and exploitation, and a call for collective political action" (2017, 56). Prunetti engages with the patterns mentioned previously in a fashion that is both personal and political. This is underscored in the following remarks: "The factory offers the blackmail of bread and exacts the right to pollute" (2012, 76; Lo stabilimento dà il ricatto del pane e pretende il diritto di inquinare); "External workers work more and are paid less, they are a more easily exploitable commodity, they accept overwork and never say 'no'" (84; I lavoratori esterni lavorano di più e si pagano di meno, sono una merce che si sfrutta più facilmente, che accetta gli straordinari senza mai dire 'no'); and "The sentence states only that Renato was exposed to asbestos, it does not say that asbestos killed him" (125; la sentenza afferma soltanto che Renato è stato esposto all'amianto, non che l'amianto l'ha ammazzato). These excerpts also illustrate how the novel's narrator, combining pain, anger, and class resentment, critiques the hypocrisies, irresponsibility, and greed of the whole establishment "from the small owner to the top of the Italian industrial class" (126; dal padroncino fino ai vertici della classe industrial italiana).

At this point, it seems pertinent to include some considerations on *Amianto*'s alleged inscription in feminist practice, which will involve addressing its similarities with and differences from what Alaimo calls the genre of the environmental "material memoir," where "the material world becomes the very substance of self" (2010, 95). On the one hand, *Amianto* unquestionably fits such a profile. Just think, for example, about how Renato "used to breathe zinc and lead, until he had a good amount of Mendeleev's periodic table tattooed on his lungs" (Prunetti 2012, 114; respirava zinco e piombo, fino a tatuarsi un bel pezzo della tavola degli elementi di Mendeeleev nei polmoni) or about how the novel works as a "counter-memory . . . mixing family history . . . personal narratives; social, political, and economic critiques; and scientific explanations" (Alaimo 2010, 96). However, some of the text's characteristics indicate that "feminism's long history of attending to body politics" (95) is not the only genealogy here. Rather, it is complemented by a tradition of peculiarly Italian, regional literature and culture interested in exploring and resisting the harmful effects of industrial labor and, by extension, capitalism itself.

To highlight some of the ways in which Prunetti's text adds to (or subtracts from) its American counterparts discussed by Alaimo, one may observe how *Amianto*'s male narrator does not display the "profound sense of uncertainty" (Alaimo 2010, 93) or dramatize the "miasma of uncertainties and interconnections in risk society" (95) typical of "material memories." In spite of and against official legal and medical discourses that, for obvious reasons, question the links between asbestos exposure and the rise of mesothelioma, the narrator is confident in his assertions, even when "engaging with scientific accounts" (93). As he puts it,

> The CAT scan is a punch in the face . . . it is easier for the health system to list as carcinogenic a cigarette one chooses to smoke rather than a manufacturing process which is imposed and endured. . . . The drugs and therapies are poisoning his [Renato's] body as much as his tumor, as much as asbestos is doing.
>
> La Tac è un cazzotto in faccia . . . più facile per il sistema sanitario indicare come cancerogena una sigaretta scelta individualmente che un processo produttivo imposto e subito. . . . Farmaci e terapie gli avvelenano il corpo quanto il tumore, quanto l'amianto. (Prunetti 2012, 99, 101)

Occasionally, Prunetti adds humor to this tragic story, recalling entertaining episodes from his father's life, interspersing his narration with vernacular expressions and scatological references, and imaginatively merging Renato first with a "metalcowboy" (2012, 143) and then, in the last chapter ("Come Steve McQueen" [Like Steve McQueen]), with an iconic Hollywood rebel—strong and vulnerable at the same time.[21] One thus wonders if his use of the Tuscan vernacular might be another way to counter the official jargons of science, medicine, and law. More generally, I wonder if humor in general and Prunetti's own politically incorrect, historically very leftist "hybris livornese" (99) specifically may be instrumental to the creation of those "alternative systems of meaning" (Alaimo 2010, 199) that texts centered on environmental-material memories ultimately strive to achieve.[22]

That said, since Renato's exposure to asbestos occurred throughout his career at many industrial sites throughout Italy, Prunetti does not focus on the asbestos-induced toxicity of a single place or on the relationship between the substance and the development of his father's disease. However, in a conversation with his mother (Francesca) that takes place when both are back home from the hospital, after witnessing the last hours of Renato's agonizing death, one specific location does emerge as an ur-site where "everything began" (2012, 109; è cominciato tutto lì). This is Casale Monferrato in Piedmont, also known as "the white town" and "the worst place in Italy, and perhaps in Europe for exposure to asbestos" (108; posto peggiore per l'esposizione dell'amianto in Italia, forse in Europa). In the following section, I thus leave Tuscany behind in order to give a fuller picture of this wounded place and of the asbestos stories it fosters.

To begin with a few historical notes, let us remember that from 1906 until its closure in 1986, the infamous Eternit industrial plant was headquartered in the working-class Ronzone neighborhood in Casale Monferrato. As I mentioned at the beginning of this chapter, the Solvay plant along the Tuscan coast literally "made" Rosignano by luring both workers and managers to live there with the promise of better pay and social distinction. Eternit followed a similar strategy along the banks of the river Po. Besides Turin, Casale Monferrato became Piedmont's most important factory town.[23] But Eternit's main product was far more dangerous than Solvay's sodium hydroxide; besides asbestos-laced roof coverings and pipes, Eternit manufactured fiber cement that, as its name suggests, was obtained by mixing asbestos dust and cement. As journalists Silvana Mossano and Michele Brambilla (2012) write, initially this material was not particularly well received in Italy. It was only after the Messina earthquake in

1908 that fiber cement boomed thanks to its ability to be produced at a low cost. Soon it became known as a sort of reinforced concrete for the poor.

As a consequence, "in Casale, asbestos was everywhere—in roofs, houses, storage places, in factories, in the air, and finally in the lungs of the population" (Iovino 2016a, 152). This diffused presence caused the death of more than three thousand of its inhabitants (and still counting). In 2014, Turin's court of appeal acquitted Eternit's billionaire Swiss-Belgian executives of environmental disaster. This controversial verdict in favor of a multinational corporation had a global impact, giving international significance and resonance to a local environmental and legal disaster. As the court sent out a message of impunity for committing this kind of ecocrime, it implicitly legitimized not only Italian productive strategies related to asbestos but also those in other countries where the substance is still widely used (i.e., India, China, Brazil, and the United States).

On a more positive note, this case was both preceded and followed by several artistic and cultural initiatives that, together with the efforts of local associations and citizen groups, contributed to the recovery and redemption of the town. Eventually, these organizations' efforts forced the Italian judicial system to revise and pass stricter laws on ecocrimes in May 2015. In addition to personal material testimonies such as Prunetti's *Amianto*, other expressive forms of storytelling and art-making focused on the communal responses to Casale's toxic history. These truth- and justice-seeking works played an important role in the city's eventual vindication and, as Gary Snyder would say, in the process of its narrative "re-inhabitation" (1995, 190).

Unafraid of tackling difficult and controversial issues, two figures in particular represent an ethically informed and socially conscious "journalistic storytelling" (Lindgren and Phillips 2016, 162). The first is Silvana Mossano in her *Malapolvere: Una città si ribella ai "signori" dell'amianto* (Evil dust: A city rebels against the asbestos "masters"; 2010) and, coauthored with Michele Brambilla, *Morire d'amianto: Il caso Eternit; La fabbrica, le vittime, la giustizia* (Dying by asbestos: The Eternit case; Plant, victims, justice; 2012). The other is Rosi Battaglia, a self-described freelance "civic editor," blogger, and documentary maker who in 2013 founded the influential online platform Reactive Citizens (Cittadini Reattivi) and the cultural project Resilient Stories (Storie Resilienti). Both writers illustrate the entanglements among health, industrial capitalism, and the environment, and both have been particularly effective in sharing information and raising public awareness about Casale's fight against asbestos. Additionally, both of them often exploit crowd-mapping to gather reports and stories from variously contaminated territories in Italy.[24]

This kind of authoritative and socially conscious journalistic work may start with an embryonic online presence that is then further developed on the printed page, on the screen, or on the stage. Its adoption of storytelling for the purpose of gathering a repository of collective memories and "inject[ing] them in the public domain" (Lindgren and Phillips 2016, 160) makes it particularly effective in stimulating a strong moral and performative (i.e., militant, activist) reaction among its audience together with an aesthetic-cognitive one (see Jansen and Urban 2017, 8).[25] Thus I approach Mossano's *Malapolvere* in light of some of the observations Jansen and Urban make about transmediality and the work of journalist Giuseppe Fava.[26] Fava's intellectual activity, they observe, "was aimed at establishing a symbiotic relationship between an ethical conception of journalism and his theatrical and narrative production" and was able to effortlessly "shift between journalism, theatre and fiction" (2017, 7).[27]

The similarly hybrid nature of Mossano's work is such that *Malapolvere* begins with a short fiction "Il Sospetto" (The suspicion), a first-person narration of an "everywoman" from Casale Monferrato who struggles to manage her fear, anger, and sadness after being diagnosed with an ominous "light pleural effusion" (2010, 14; leggero versamento pleurico)—possibly a first sign of mesothelioma. The next section includes the actual testimonies of real local women who either got sick themselves or had family members who were "somehow swept away by the dust" (11; travolte, in qualche modo, dalla polvere). The work concludes with a historical chronicle of the socioeconomic costs of Eternit's presence in town. Mossano thus combines the rhetorical power of fiction and nonfiction for maximum reach and effect.[28]

Unlike Fava, Mossano does not experiment with theater herself. However, her text continued its mediatic dispersion when it inspired actress and playwright Laura Curino to adapt it into a stage monologue entitled "Malapolvere: Veleni e antidoti per l'invisibile" (Evil dust: Poisons and antidotes to the invisible; 2012). An established artist and theater personality—together with other popular Italian "narr-actors" such as Dario Fo and, more recently, Marco Paolini, Marco Baliani, and Ascanio Celestini—Curino represents the tradition of a socially committed narrative theater focused on topics of civil relevance and public interest.[29] Taking advantage of the emotional impact of oral storytelling, this kind of theatrical production establishes a close relationship between the performer and the audience and, by boosting the level of collective engagement with and awareness of the addressed topic, "expand[s] the act of narration into that of cultural activism" (Jansen 2017, 23).

In Curino's captivating ecomonologue, she retraces the geo-history of Casale from ancient times to the present, interspersing her factual narration with voices from several nonhuman entities: a speck of dust, a tree, one of the bicycles used by the workers to reach the Eternit plant, the polluted waters of the river Po and two of its tributaries (Lanza and Mellana), a dam, fog, hail, the city Tower, a fountain, the bronze equestrian statue in Casale's main square, the city Castle, and an old book. In this haunting rendition, these things are transformed from silent witnesses into active participants in an ongoing conversation. Through the decades, they have impotently observed the white-dusting of the town and its surroundings and the formation of the deathscape that followed. Now they are able to communicate both the absolute transcorporeal pervasiveness of asbestos dust and the message that any successful project of regeneration and reshaping of our environmental imagination should heed the collective and coemerging stories of people, places, and things. Only these entangled narratives are the "antidotes to the invisible"—able to cure and enlighten at the same time. And in this sense, their task will be unsurprisingly similar to the one emblematized by the bronze horse in the main square of Casale, which, as we learn toward the end of Curino's piece, plans one day to tear itself away from its pedestal, run to the sea "until his coat will smoke," and, after finding a narwhal tooth on the beach, become a powerful "magical unicorn able to heal and console everything and everyone" (Curino 2012).

Given recent developments in Casale Monferrato, there is good reason to believe that this imaginary unicorn is arguably doing its job and, out of metaphor, that such a narrative healing and consoling process, integral to collective mobilization and the achievement of any socioenvironmental justice, is already underway. For example, in the fall of 2016—ostensibly also because of the combined impact of the "asbestos stories" discussed in the previous pages—the reclaimed Eternot public park took the place of the Eternit plant as a memorial site that affirms the city's collective values. Its once contaminated grounds are now sites of initiatives and alternative landscape practices such as the Eternot Vivarium, a permanent "live monument" designed by artist Gea Casolaro. By replacing the volatile asbestos fibers with a cultivation of *Davidia involucrata* (dove tree)—a humble, widely diffused plant with peculiar white leaves and flowers—the monument reanimates this wasteland and provides a powerful material and symbolic representation of Casale's fight against Eternit. In short, it sows both hope and resistance, "reseeding the grounds of the past for the materialization of a different future" (Casid 2005, xiii).[30] Additionally, as a reminder that asbestos was never solely confined to this place but always had a global impact, every April on International Workers' Memorial Day, which remembers those killed at work or by diseases contracted at work, dove tree saplings are sent to persons and institutions throughout the world that have distinguished themselves in the ongoing struggle against asbestos contamination. In this regard, I cannot help but think of this practice in the light of Jill Casid's emancipatory notion of "nomadic garden[s] . . . as a tactics for making and transforming" (2005, xv).

In an even more ambitious objective, Casale Monferrato plans to continue its efforts toward decontamination and cultural recovery and become the first asbestos-free town in Italy and Europe. Since any environmental crisis points to a fundamental cultural crisis, it is only natural that a project of cultural rejuvenation is accompanied by one of territorial reclamation and reinhabitation. Indeed, such objectives corroborate the notions that diseased cultures and natures should be recovered and remediated at the same time; that the health of bodies, minds, and landscapes cannot but go hand in hand; and that "conceptual revolution(s) that allow us to see all . . . together" (Iovino 2016a, 154) and envision alternative environmental imaginaries are not just unicorns on a beach.[31]

5
Concrete and Asphalt
Geographies of Environmental Disruption in Modern Italy

Concrete, after fossil fuel, is the most environmentally destructive substance we have.

—Amitav Ghosh, "From Bombay to Canton: Traveling the Indian Ocean Opium Route"

Earth and flesh almost mingle and the body turns into landscape and the landscape is embodied.

Terra e carne quasi si confondono e il corpo si fa paesaggio e il paesaggio prende corpo.

—Franco Arminio, *Terracarne: Viaggio nei paesi invisibili e nei paesi giganti del Sud Italia*

Among the many intriguing artworks displayed at the *End of the World* exhibition at the Luigi Pecci Centre for Contemporary Art in Prato (October 16, 2016–March 19, 2017), Julius von Bismarck and Julian Charrière's *Clockwork* (2014) stands out as an appropriate visual anchor and entry point into this chapter. It eerily displays twelve rotating concrete mixers arranged in a circle in a large room, as if each of them represents one hour in a very special clock that, rather than indicating time, marks a progressive dematerialization of space. If, indeed, "landscapes are assemblages constituted by humans and nonhumans, material and semiotic processes, histories both real and partially remembered"

(Ogden 2011, 35), this piece that speaks of concrete's potential to constitute, transform, and destroy the reality we inhabit could hardly be more evocative. As the two artists explain, these tools "form an 'erosion machine' designed to accelerate the decay of Vienna. The mixing drums are filled with architectural debris taken from various buildings across the city. The rotation of the mixers transforms these man-made angular bricks into round, organic forms. Over a long period of time and through continuous impact, the bricks are turned into pebbles and ultimately become powder. Inside the exhibition space this process is tangible through noise, clouds of dust, and vibration" (von Bismark and Charrière 2014).

Ironically subverting the standard function of concrete mixers to prepare mortar or cement that shall eventually solidify and fill a space, *Clockwork* evokes a sort of creation in reverse—one that, depending on the viewer's perspective, reminds one of either an ecological utopia or an apocalyptic dystopia. As a utopia, it shifts our attention from the (over)built to the unbuilt, from our contemporary excesses and *horror vacui* to a primeval emptiness, an absence, and a tabula rasa from which to start over, ideally in more environmentally friendly ways. As such, it points to the erosive effects of time on things, places, and bodies, both living (organic) and inert (inorganic). It plays with the idea that the same machinery and technology used to construct our urban spaces and infrastructures could creatively function in reverse to speed up their deconstruction and dematerialization. Finally, it reminds us how all these processes bear on our senses and bodies.

At the same time, the slow metamorphosis of Vienna—a city that many consider among the most beautiful, livable, and cultured places in Europe—into a powdery substance may refer to the progressive transformation of this urban environment due mostly (even though not exclusively) to concrete and the irrational overbuilding of what is generally perceived as a "good place" into a nonplace. *Clockwork* invites one to think about other metropolitan areas in Europe and beyond that, over the last few decades, have been characterized by simultaneous expansion and fragmentation, a literal and metaphorical loss of center through pulvuscular sprawls of asphalt and cement. Thus when writer Vitaliano Trevisan describes the soul of the Italian northeast with the image of a "giant concrete-mixer" (2010, 18; gigantesca betoniera), he sounds very much in sync with what *Clockwork* is communicating.[1]

Incidentally, these are the sort of "concrete" processes that, according to Milanese architect and urbanist Stefano Boeri, have been turning cities into what he calls "anticittà" (anticities)—that is, places characterized by homogeneity, individualism, and fragmentation "constituted by a myriad of monocultural islands disinterested in the functioning of the geographic and anthropologic organism they nonetheless belong to" (2011, xiv; composte da una moltitudine di isole monoculturali e disinteressate al funzionamento dell'organismo geografico e antropologico a cui pur appartengono). Reading asphalt and concrete's impact in light of the notions of utopia and dystopia and considering how they affect, as both materials and tropes, the production of some Italian spaces is thus one of the goals of the next section.

A Plunge into the "Betoniera del Nord-Est": From Expo 2015 to Giorgio Falco's "Cortesforza"

In *Gli iconemi: Storia e memoria del paesaggio* (Iconemi: History and memory of landscape; 2001), geographer Eugenio Turri, in collaboration with photographer Mimmo Jodice, provides a poignant visual and narrative map of the Lombard Po Valley, recording its drastic transformations in recent years. A few years earlier, in his *Semiologia del paesaggio italiano* (Semiology of the Italian landscape; 1979), he had already observed that there is often a "correlation . . . between the large quarry that cuts open a mountain side, the nearby cement factory and the recently constituted, not-far away urbanized area" (1979, 26; correlazione . . . tra la grande cava aperta sul fianco di un monte, col vicino cementificio, e la non lontana area urbanizzata di recente costituzione). More recently, the report released by the Istituto Superiore per la Protezione e la Ricerca Ambientale (ISPRA; Higher Institute for Environmental Protection and Research) in June 2017 states that "Lombardy . . . confirms its status as the Italian region with the highest absolute values and percentages of consumed soil" (2017, 63; La Lombardia . . . si conferma la regione italiana con i valori assoluti e percentuali più alti di suolo consumato). Inspired by Turri and Jodice but also by David Harvey, another geographer who pays particular attention to the transformative spatial dynamics of modernity and postmodernity, and Ursula Heise, who has considered utopia from an environmental perspective, I reflect on the active roles concrete and asphalt play within the current dynamics of space occupation and production.

In his seminal book *Spaces of Hope*, Marxist geographer David Harvey reevaluates utopian thinking as a means of counteracting the dismal global effects of capitalism's ability to constantly reinvent itself and ultimately advocates for "a more robust utopianism, a 'utopian dialectics,' grounded in real historical and geographical tendencies" (2000, 195). In particular, in the chapter "The Spaces of Utopia," Harvey tells us how his troubled hometown, Baltimore, exemplifies in bricks and concrete his previously elaborated theory of "uneven geographical development" (148)—that is, the ways in which capitalism has managed to occupy and produce space despite criticism and resistance. As Harvey lists some of this distressed city's foremost socioeconomic and environmental problems—chronic poverty, lack of access to good education, "deindustrialization, the abandonment of the inner city, and the concomitant investment in privatopic development for the privileged" (138)—he also contextualizes Baltimore in larger regional, national, and global frameworks so that one is encouraged to trace and recognize similar dynamics elsewhere in the world.

More recently, in an essay entitled "The Environmental Humanities and the Futures of the Human" that could dialogue with and add Harvey's text among its genealogical referents, Ursula Heise concludes her reflections by encouraging environmental humanists to analyze critically not only declensionist and dystopian narratives but also utopian ones that, in her view, offer "more constructive visions of the long-term future" (Heise 2016, 30). At the origin of Heise's invitation is her concern that apocalyptic narratives "seldom bother with how to generate and sustain more functional social structures and healthier natural environments" (30) and have become so prevalent in environmentalist discourses and practices that their "political power has become blunted" (29–30). Heise, however, is also quick to note that there are serious risks associated with utopian environmental narratives. In some instances, in fact, utopian thinking can be an instrument for resistance and liberation that allows us to reimagine ontological borders and the limits of anthropocentrism. In other cases, however, utopian projects are complicit with techno-scientific ideas of progress and development and, in turn, tendentially authoritarian, corporative, neoliberal interests.

Incidentally, Harvey had already observed how many utopian schemes have reproduced the authoritarian tendencies of capitalism in seeking to impose spatial homogeneity on diverse populations. To illustrate her point, Heise refers to the (utopian) ecomodernist strategic vision of the San Francisco–based Breakthrough Institute that is optimistically centered on human ingenuity and, as its website states, aims at "identify[ing] and promot[ing] technological solutions to environmental and human development challenges" (Breakthrough Institute, n.d.). Namely, under the banner of a "good Anthropocene," the authors of the Breakthrough-related "Ecomodernist Manifesto" invoke science, technology, bioengineering, and terraforming practices as the best means to address the complex environmental challenges of the twenty-first century (Nordhaus and Shellenberger 2015). The problem with this kind of utopian position, Heise (2016) observes, is that it is unaware of and unconcerned with the large contributions of the environmental humanities—not to mention that, as critics of the notion of a "good" Anthropocene point out, it is also founded on "bad science" (Hamilton 2015, 236).

As I pondered how some of the insightful points raised by Turri, Harvey, and Heise might resonate with and apply to specific places and discourses in Italy, the "progress-oriented" (simultaneously dangerous, in Heise's [2016] sense, and degenerate, according to Harvey [2000]) utopian narrative that came to my mind was the one informing Expo 2015, the World Expo held in suburban Milan. Expo 2015's relevance to cement and the rest of my discussion becomes immediately clear when we recall with Daniela Danna that "the land transformation in the Rho-Pero-Milan trapezoid started with the preparation of the terrain, then cement was poured. . . . The terrain was leveled, local streams were canalized, and the first artifact, the *piastra*, was built. The latter is a concrete platform, about 2 km long and between 350 and 750 meters wide, that is connected to the drainage system, water supply, and electricity grid, with about 10 km of roads" (2017, 911–12). Literally a "blanket of a becoming" (Thompson 2019, 42), as this concrete slab materialized, it transformed the territory and the lives of its inhabitants. On the one hand, it homogenized and flattened differences; on the other, it separated and complicated a number of previously existing dynamics, from water and energy flows to human and animal ones.

Centered on the theme "Feeding the Planet, Energy for Life," Expo 2015 projected faith in human capacity and the future. Echoing the position of the ecomodernists, its various organizers and corporate sponsors also advocated for the importance of scientific research and agro-industrial innovation as crucial solutions to world hunger, therefore obtaining that magic (though rarely achieved) condition of sustainability. In simple terms, the Expo's core message and underlining narrative can be summarized as follows: the fundamental cure to malnutrition and the prevention of alimentary-related illnesses (and thus the improvement of socioeconomic conditions all over the globe) is linked to our capacity to produce and market more food. This objective might be obtainable with the help of new and developing technologies (e.g., genetically modified organisms, biotech), targeted financial operations, and further industrialization of the food chain. In addition, by acquiring and putting to good use swaths of unproductive land—such as those that can still be found in Africa, Asia, and Latin America—crops can be grown for human consumption and for allegedly ecofriendly biofuels (to help offset our ever-increasing energy demands).

So what is the problem with this optimistic scenario of abundance and apparent sustainable growth? As it turns out, there are several causes for concern. Just to mention the most salient, the notion of sustainability itself contains many dangers. While delineating his (utopian) vision for a new self-sustainable development based on the valorization of local traditions and heritages, for example, Italian urbanist Alberto Magnaghi observes that sustainability "often covers up the structural causes of social and environmental degradation through actions supporting the established development pattern, without ever calling into question the rules generating it" (2005, 34).[2] More recently, Walter Mignolo echoes and amplifies Magnaghi's reservations in an article that distinguishes between and juxtaposes the notions of "sustainable development" and "sustainable economies." Mignolo argues that sustainable economies should be privileged, since, unlike sustainable development, they do not simply view economic growth as the commanding principle; rather, they "are articulated in the vocabulary and philosophy of harmony, reciprocity, and communal, non-hierarchical organizations" (2016, 4n3). As he puts it, while sustainable economies could regenerate a communal, social fabric of hope, sustainable development "could solve the ecological problem of sustainability, but it cannot solve most of the problems existing today, such as inequalities, interstate conflicts, migrations, and refugees" (23). What follows is that the kind of sustainable utopian narrative associated with Expo 2015, with its focus on quantity and production, is essentially blind to the social causes that lie behind the global problem of hunger, as this narrative is fundamentally uninterested in spreading a different alimentary culture, creating a new pattern of food distribution, or preventing the repetition of neocolonial, environmentally deleterious practices such as land-grabbing, displacing local populations, and establishing monoculture plantations. To put it in Mignolo's terms, it speaks the language of development and constant growth more than anything else.

But what is even more relevant for this discussion is that the contradictions, tensions, and limitations emerging from the Expo's utopian discourse at the global level may become more tangible and visible if one underscores also its concrete material implications and situated ecological impact at the local one. Especially if observed through the lens of Harvey's notions of "utopianism of spatial form" and "degenerate utopia" (2000, 167), Expo 2015 constitutes a useful case study that not only highlights the shortcomings and contradictions of the neoliberal project but also shows how cement and asphalt are closely imbricated with such a project—or to put it differently, how central these substances are to the present phase of Italian capitalism.

Spread over 104 hectares (approximately 257 acres) in an area covering the jurisdiction of three different Lombard municipalities (Milan, Pero, and Bollate), Expo 2015 is considered one of Italy's major public infrastructural projects (a so-called *grande opera* or megaevent) and among the most ambitious urban development enterprises in recent years. As the existing literature illustrates, its advocates and managing companies imagined and marketed it over the four years it took to complete as a "zero-impact" project. Expo 2015 would involve no substantial new construction. This was miraculous for a country that, as Antonio Cederna states in his *Brandelli d'Italia* (Shreds of Italy), was "among the major producers and consumers of cement in the world, two or three times that of the United States, Japan and Soviet Union: 800 kilos for each Italian" (1991, 10) and that Salvatore Settis describes as one in which real economic and sociocultural development, from the 1950s onward, has been often blocked rather than supported or sustained by state-sanctioned "cement floods" (2010, 82; alluvioni di cemento). At least on paper, Expo 2015 organizers planned to recuperate, reutilize, and valorize existing urban infrastructure in the chosen territory.[3] However, as mentioned earlier, some new construction was eventually needed to improve the viability of the area. In the end, approximately two hundred kilometers (125 miles) of new streets and highways, a new water canal connected to the historical Navigli, and a new metro line were deemed necessary because they allegedly constituted a substantial improvement for the local communities and all the inhabitants of the Province of Milan.

Expo 2015 would also discourage criminal infiltrations and speculative operations, as planners arranged to have it partially funded by money seized from criminal organizations, thus sending a message of legality and transparency to potential bidders and investors. To complete this ideal picture of livability and ecological balance between the old, historic landscape and the new one, mobility and quality-of-life issues for the suburban and urban areas involved would be solved by adding extensive walking and bicycle routes. Only renewable energies would be used, and large gardens would provide the city with seasonal, organic vegetables and fruit. In short, the exposition's model citadel—a sort of upgraded, twenty-first-century "city of the sun" (or perhaps, more appropriately, "city of the fog," given its latitude and longitude)—would be an icon representing the vital energy, progressive modernity, and technological achievements of Milan, Lombardy, and by extension, all of Northern Italy. In short, it would tell the world about the nation's capacity to build new infrastructure in which functionality, aesthetic appeal, and sustainability come flawlessly together.

The two main landmarks located at the center of Expo 2015 would be powerful embodiments of such aspirations. Despite reminding some of a new Altar to the Fatherland,[4] the first of these, the eye-catching Italian Pavilion (Padiglione Italia), was said to be inspired by an urban forest. According to the organizers' description, it features "a primitive but also technological design which envelops the building" and displays a "webbing of branches [that] produces an alternation of shadow and light that alludes to a Land Art sculpture" (Expo Milano 2015, 320). The second, the *Tree of Life*, an enlightened sculpture (in the sense that, like a Christmas tree, its lights could be switched on and off), was made of steel and wood and was designed to symbolize "primordial nature, the great power from which everything was generated" (320). This monument, with its message of ecological sustainability and resilience, may appear in a different light if one thinks about the sources and hidden environmental costs of its constituting materials. Far from being natural or primordial and rather implying a familiar attitude of remaking and improving nature, the *Tree of Life* cannot but remind us of the scarcity of large forests in Italy lamented by F. T. Marinetti in chapter 1, of the country's dependency on oil as mentioned in chapter 2, and of the high toxicity of steel production discussed in chapter 4.

Shifting our focus to the Italian Pavilion, Italcementi, the company responsible for its design and construction, proudly reveals that the innovative biodynamic concrete used to build this zero-emissions architectural marvel not only is made of 80 percent "recycled aggregate partly coming from the machining scrap of Carrara marble" but also is able to reproduce the pattern of a brilliant white "petrified forest." Moreover, it promises to miraculously "free the atmosphere from smog," thus "contribut[ing] to the improvement of air quality" (Italcementi 2015). Without necessarily questioning the aesthetic appeal of this building or the benefits of this pure, futuristic construction material, let us underline here not only how greenwashed these statements are but also how they blur the line between utopia and dystopia. In other words, they display the same kind of clashing logic informing the entire Expo 2015 operation—namely, as Wu Ming 2 states in his *Il sentiero luminoso* (The shining path), of "covering with cement an agricultural area six hundred times larger than Milan's Piazza Duomo under the slogan 'Feeding the planet'" (2016, 259; coprire di cemento una superficie agricola pari a seicento volte Piazza Duomo e piantarci sopra lo slogan "Nutrire il pianeta") and, similarly, of arguing for the preservation of water as a communal resource and universal right by altering with concrete and asphalt the nature and course of existing water channels.[5] In this sense, the biodynamic concrete of the Italian Pavilion is a quintessential "vibrant matter"—potentially vital and agentic in the sense discussed by Jane Bennett. Like the powdered bricks in *Clockwork*, it emblematizes the coexistence of antipodal notions of constructivity and collapse. Although this substance may well "generate and sustain more functional social structures and healthier environments" (Heise 2016, 30), it is easy to envision how its widespread presence and utilization would negatively affect myriads of now completely superfluous living trees, not to mention the whole ecosystem of the Apuan Alps, where, as I noted in chapter 3, the scrap of Carrara marble originates.

Perhaps unsurprisingly, the Environmental Justice Atlas (EJAtlas) still lists Expo 2015 among the current cases of major environmental conflicts in Italy. However, as the EJAtlas website illustrates the range of its environmental, health, and socioeconomic impacts (from air pollution and deforestation to militarization, loss of landscape, and erasure of sense of place),[6] it also reminds us about some of the counterutopias envisioned by many groups (EJAtlas 2015). Local citizens and members of environmental justice organizations and social movements all mobilized by occupying buildings and public spaces, participating in street marches and blockades, organizing guerilla theater, and painting murals to prevent Expo 2015 from happening and, at the same time, to suggest alternative—or, if you wish, "dialectical" (in Harvey's sense)—options.[7] Although these groups were able to force a revision of the Waterways/Darsena project, their efforts were only partially successful. However, as Rebecca Solnit posits in her *A Paradise Built in Hell*, disasters provide glimpses of not only the apocalypse but also utopia. They do not merely destroy but create: "Disasters provide an extraordinary window into social desire and possibility" (Solnit 2009, 6).

In this sense, some of Gerolamo's reflections on Expo 2015 in Wu Ming 2's *Il sentiero luminoso* resonate with Solnit's insight, as they indicate both that what is now largely perceived as another environmentally degraded area may offer a chance for a future new beginning and that forms of resistance are still very much present even after the conclusion of this event:

> An isolated house, suffocated by roads, hurdles and construction areas, gifted us with the classical image of resilience to disaster. The immense empty parking lots suggested visions and utopias, like that of transforming them in endless skating rinks. A thin layer of water and the right temperature would be enough, and they would become the crown-jewel, the largest in Europe, a wonderful example of reuse and energy for life.
>
> Una casa isolata, stretta nelle spire di strade, transenne e cantieri ci ha regalato la classica immagine di resistenza al disastro. Gli enormi parcheggi vuoti ci hanno suggerito visioni e utopie, come quella di trasformarli in sterminate piste di pattinaggio, basta un velo d'acqua e la giusta temperatura, ma ci siamo detti che subito diventerebbero un fiore all'occhiello, le più grandi d'Europa, un bellissimo esempio di riutilizzo ed energia per la vita. (Wu Ming 2 2016, 272)

Even though Gerolamo's "skating rinks" utopia (global warming permitting) may not find its spatial materialization in the future, these now empty lots can still be imaginatively transformed into an emblematic site where a mock ritual of protest is performed:

> I extract [the brick] from my knapsack, I lay it on the asphalt and we form a circle around it.—"Of useless cement"—I begin to chant.—"Let's get rid of it together"—the assembly responds. "Of tangential roads"—I resume.—"Let's get rid of them together."—"Of the waterway."—"Let's get rid of it together!"—"Actually," someone corrects me, "we got rid of that!"
>
> Lo estraggo dallo zaino, lo appoggio sull'asfalto e ci mettiamo in cerchio tutt'intorno.—"Dal cemento inutile"—salmodio in tono grave.—"Liberiamoci insieme"—risponde l'assemblea.—"Delle tangenziali esterne"—riprendo.—"Liberiamoci insieme."—"Dalla Via d'Acqua."—"Liberiamoci insieme!"—"Anzi no" corregge uno—"Ci siamo liberati!" (272)

The fate of the whole Expo 2015 area, similar to that of other impermanent and disposable structures elsewhere, is still not very clear. Gerolamo's modest, utopian project stands in the middle ground between the proposals for outright demolition and plans to partially reopen it to host a research center for the University of Milan, art exhibitions, and other public events.[8] Similarly, the *Tree of Life* and the biodynamic-concrete-made Italian Pavilion still stand as both revealing symbols and material texts inserted in this suburban landscape. And as the airy, perforated frame of the latter structure embodies not only the technological advances and progressive aspirations but also the problematic conceptual holes and wounds affecting Italy's love affair with concrete, it becomes one of those "places that can . . . enlighten the life of other places, of other collectives" (Iovino 2016a, 2). While still paying attention to both the mutual interrelations existing between places and human beings, and the role that imagination and narratives play as activators and recorders of environmental concerns, we will now explore one of these revealing "other [non]places" just a few miles away from the Expo 2015 site.

Lying just outside "the southwest area of Milan, along Highway 494, more precisely in the strip of land between Vermezzo and Abbiategrasso" (Falco 2009, 17; zona sud ovest della provincia di Milano, lungo la Strada Statale 494, e più precisamente nel lembo di terra tra Vermezzo e Abbiategrasso), in the "widespread periphery" (Trevisan 2010, 13; periferia diffusa[9]) of the Italian northwest, Cortesforza is the fictional and yet very realistic hinterland "municipality with 1574 inhabitants" (Falco 2009, 17; comune di 1574 abitanti) that provides the backdrop for the nine short stories of ordinary madness in Giorgio Falco's award-winning collection *L'ubicazione del bene* (The site of goodness; 2009).[10] As Falco's title indicates, the central issue here is philosophical as well as material, moral, and topographic and bureaucratic, since it evokes the impossible aspirations and contradictions at the heart of any utopia—namely, locating goodness and realizing it in material form (*bene* as a noun in Italian means both "good" and "asset" or "possession").

Cortesforza is fundamentally similar in structure and concept to the real sprawling residential developments situated among industrial sheds, megacommercial centers, and discount outlets that, from the seventies on, have popped up in suburban areas around Milan, in the Po Valley, and all over Lombardy and the Veneto region. At the time of their establishment, they represented an escapist desire for isolation and protection from the dangers of the city. At the same time, Cortesforza is not the by-product of criminal real estate speculations and illegal practices of abusiveness (*abusivismo*) like those that, for example, distinguish the even more utopian-sounding New Dawn Village in

Simona Vinci's novel *Rovina* (Ruin; 2007), which is set in the bordering Emilia-Romagna.[11] Additionally, even though Cortesforza is situated in one of the most polluted areas in Europe—sharing the honor with Germany's Ruhr valley, with its dangerous concentrations of atmospheric particulate matter and dust—it is not the site of a specific toxic spill or environmental catastrophe, like the nearby town of Seveso.[12] Nonetheless, as Falco's stories make clear, this place, with the kind of dwelling it entails and lifestyle choices it encourages—or better, imposes—is far from being a healthy model of "an inhabited site built on a human scale" (Falco 2009, 17; un centro abitato a misura d'uomo) or a sign of enlightened environmental planning or ecologically sensible and sustainable politics.[13]

On the contrary, this development is similar to what Harvey calls "privatopias" and modern-day "degenerate utopias"—that is to say, "happy, harmonious, and non-conflictual" spaces designed to shield inhabitants from the real world in order to "soothe and mollify, to entertain, to invent history and to cultivate nostalgia for some mythical past" (2000, 164). Although it bears an allusively ancient, aristocratic name that is perfect for a real estate market that is always eager to re-create historical authenticity, Cortesforza is just another "area of florid ruin" (Falco 2009, 47; zona di florida rovina).[14] It is one more "non-place" (Augé 1995) that, in its structural inability to shape a meaningful and reassuring sense of community, actually hides in plain sight a profound sense of malaise, alienation, and isolation, and ultimately, it cannot but reveal the "laceration of our environmental and anthropological fabric" (Ferroni 2010, 76).

Mirroring the sclerosis that, as Harvey (2000) reminds us, is peculiar to most utopias that are unable to deal with temporal change, Cortesforza is characterized by a painful immobility that weighs on things and people, while everything, despite being new and newly built, is already affected by a sense of erosion and consumption. Falco's stories take place behind domestic walls, shaped hedges, mowed gardens, extensive playgrounds, designer pets, plastic fences, rooms "filled with still wrapped, piled up commodities" (Falco 2009, 86; merce accatastata e ancora imballata), and other deceptive exterior signs of consumeristic well-being. They consistently address the by now familiar notion that one's physical and psychological health (or the lack thereof) is embodied in the structure of the landscape and, in turn, points to the inextricable interrelations between outside and inside, places and bodies.

Given the fundamental interdependence between human beings and their surrounding environment, it is worth considering how this particular expanse of concrete, asphalt, and bricks that was thoughtlessly situated in an empty field next to a highway exit interacts and shapes the minds and bodies of its inhabitants.[15] What is troubling in Cortesforza's own "mind"[16] and that of its concrete clones in the "entire world" (Falco 2009, 106; mondo intero)? How does one live in spatially homogeneous and undifferentiated (non)places where people "find it sensible to drive twenty kilometers to buy six bottles of bubbly mineral water" (106; trova sensato fare venti chilometri in macchina per comprare sei bottiglie di acqua minerale gassata) and "a thirty-nine-year-old man" (un uomo di trentanove anni) risks being considered a maniac if he walks alone or "sits in a communal garden . . . with a book in his hands" (135; in un parchetto . . . con un libro in mano)?

A closer look at some of Falco's texts and, in particular, at the representation of (trans)corporeality and the human-animal relationship may give us some answers. Parallel to the expropriated and violated agricultural landscape that human beings now occupy is their relationship with the pets they think they desire and the pests they loathe. But our perception of these categories is blurred as, time after time, the animals are fenced in, caged, poisoned, or squashed under the wheels of passing cars, merging with the asphalt until they become indistinguishable from it.[17] As if they were "snarge" themselves—that is, "collisions between forms of fossil fuel–based human mobility and solar-based animal mobility" (Kroll 2018, 82)—human beings here equally fail, collapse, and die.[18] They are unable to understand both the continuity between the natural and the cultural (or the "constructed") and their own biological finitude: "Nobody among the new residents thinks that death is imminent" (Nessuno dei nuovi residenti pensa che la morte sia imminente) but merely something that "looks like an intrusive anomaly" (Falco 2009, 11; sembra un'anomalia abusiva).[19]

In this part of the world, lives, relations, friendships, things, marriages, animals—all become disposable items, especially when they are not perceived as fitting within a virtual, digitally enhanced, and above all, marketable "perception of places" (Falco 2009, 99; percezione dei luoghi). Everything ends up like the "rubble, iron, wreckage, concrete, ruin" (131; macerie, ferro, cemento, calcinacci, tegole, pezzi di mattoni) produced by the construction process and illegally discarded into the roadside ditches just outside the residential area. In other words, in Cortesforza—a consumer-oriented, upper-middle-class suburban wasteland designed to represent the concretization and spatial realization of neoliberal projects of perennial expansion, efficiency, and feasibility, perfused by a stereotypically Northern "rhetoric of doing" (Varotto 2014, 119; retorica del fare)—death is just a useless, unprofitable distraction, a fake and meaningless preoccupation that, just like almost everything else, needs to be ignored and discarded.

In this environment where nature and animals constitute "a compressed, mashed presence confined to interstitial spaces or to amusement parks" (Tomasi 2014, 105), the first story, "Onde a bassa frequenza" (Low-frequency waves), begins by focusing on the many pests (rodents, roaches, pigeons) that infest the protagonist's apparently clean suburban habitat and on his decision to quit his job and start an extermination business. The poisons he uses and the low-frequency waves of the story's title, however, ultimately have more of an impact on him than on the pests. While he "feels on himself the smell of the disinfesting substance [and] his skin stinks" (Falco 2009, 6; sente su di sé l'odore del disinfestante ... gli puzza la pelle), his own life and family relations are coming progressively apart, simultaneously poisoned by magnetic waves, toxic materials, and his equally noxious, fraudulent financial behavior. As contamination spreads from animals and landscapes to the protagonist's body and then to business transactions and ethical attitudes, the collection thus begins with a tale that unequivocally hints at the various transcorporeal correlations at play throughout Falco's work.

The topic of pest infestation and extermination recurs in "Piccole formiche bianche" (Small white ants), which, while evoking quintessentially Calvinian textual landscapes, chronicles a young couple's struggle to restructure one of the few old farmhouses still standing within the newly constructed area.[20] Like "Onde a bassa frequenza," this short text also deals with the impossibility—despite any attempt to stick and conform to dualistic, anthropocentric worldviews—of drawing definitive lines between the dimensions of the natural and the civilized. Thus while masons work hard to remodel the roof and make additions to the building's existing structure, other, more powerful nonhuman agents (plants, insects, underground water, climate) are busy undermining their efforts. As the story reveals the commingling of lives, materials, elements, and forces at play in this place, the couple continues to misread the messages the local environment sends them ("They are saddened by the falling of the leaves, but they don't believe it's a sign of something" [Falco 2009, 78; Sono dispiaciuti per la caduta delle foglie ma non credono che sia il segno di qualcosa]). At the uncanny conclusion, worthy of a gothic horror tale, after the house's own wooden "arteries" have been injected with pesticide, the building's inorganic body—with its bricks, *béton*, and wooden beams—comes alive, although not in the way its new owners originally expected: "The saturated walls regurgitate a white mass, and Silvia backs a meter away" (80; I muri saturi restituiscono un rigurgito bianco, e Silvia indietreggia di un metro).

Other stories dramatize similar reactions of repulsion and fear at the thought that the natural world may somehow invade and contaminate Cortesforza's artificial "goodness" made of bricks and cement. For example, in "L'ubicazione del bene" (The site of goodness), the piece that gives the title to the whole collection, a wife (Alessandra) is horrified when she discovers a small worm curled into a head of cauliflower she purchased ("I don't want a grub in my kitchen! This is still my kitchen!" [Falco 2009, 22; Non voglio una larva in cucina! Questa è ancora la mia cucina!]), while a brand-new car owner (Gianni) is afraid that his prized vehicle will be scratched by some curious monkeys at a nearby "Zoo Safari Park"—a place where, according to him, nature should remain conveniently packaged and tamed in order to meet and satisfy human consumption.

In "Essere sul punto" (On the point), the purchase of a dog, mostly seen as a child surrogate, sets off a flea infestation that portends larger and more dangerous disintegrations, confusions between the organic and the inorganic, and material collapses:

> [The fleas] are so fast that they seem something else, he is not sure that they really are fleas, they are dark and yet transparent—he sprays—perhaps they are just invisible, jumping parts of the wall.
>
> Le pulci sono cosí veloci da sembrare altro, non è sicuro che siano davvero pulci, scure eppure trasparenti—spruzza—forse solo parti ormai invisibili del muro, che saltano. (Falco 2009, 73)

Finally, to conclude this brief overview, in "Oscar," the narrator takes us (or better, drives us, since walking is again off-limits here) to the basement of one of Cortesforza's cottages (*villette*), where the "movable paradise" (paradiso traslocabile) constituted by "a five-hundred-liter, three-meter-long Amazonian fish tank" (Falco 2009, 58; un acquario amazzonico da cinquecento litri lordi, lungo quasi tre metri) betrays its name when it becomes the site of Siamese fighting fish battling one another to the death. As their gathered owners, all career-centered individuals employed by the same corporate business, "gain clarity in [their] working relationships" (59; acquis[tano] chiarezza nei rapporti lavorativi), readers grasp the similarities between the fish and these humans.

Indeed, both live short and anguished lives behind walls made of either glass or brick, and there is a very thin layer that separates any artificial (utopian) paradise from a hellish (dystopian) existential reality.[21] Unsurprisingly, at the end, we encounter both the protagonist and his fish enclosed in the former's car, staring in equal puzzlement at their respective reflections on the windshield and on the fish tank.

As Falco's collection evokes utopia only to turn it into a dystopia and illustrates in the process the "shortcomings of contemporary [sub]urbanity" (Schliephake 2015, xxii), it reveals the distance between the economic logic informing rootless and sprawling places such as Cortesforza and any truly eco-logic and sustainable visions. Far from being built on a human (and, even less, nonhuman) scale, it is instead a quintessentially alienating and biocidal site constructed on the faulty premise of an ontological separation and dissociation between the natural and the cultural, the human and the nonhuman. And as such, it constitutes the fictional, antipodal image of the sort of place architect Stefano Boeri theorizes about in his *L'anticittà* (Anticity; 2011). One of the strongest voices against Expo 2015, Boeri not only underlines the importance of the notions of self-limitation and renaturalization, which are deemed necessary to revitalize degraded landscapes, but also advocates for a nonanthropocentric urban ethics and biopolitics capable of assigning a new value to neglected spaces. As he puts it, a "nonanthropocentric ethics does not abandon human beings; it simply repositions them at the center of a new discursive order, one in which humanity is no longer alone on life's pedestal" (2011, 109; l'etica non-antropocentrica non abbandona dunque l'uomo; semplicemente lo posiziona al centro di un nuovo ordine del discorso, dove l'umanità non è più sola sul piedistallo della vita).

Boeri's argument acquires a particular resonance after we witness how Falco's characters are completely blind to the fact that "human beings and animals are united in a similar fate" (Falco 2009, 62; uomini e animali uniti dal medesimo destino) and fail to perceive that concrete and asphalt are also part of the same "hybrid collective" (Schliephake 2015, xxxi). Boeri's alternative invitation to recognize a shared existential situation between humans and animals—to consider them as an assembled "network of actants"—and to prevent them from being disruptively placed and displaced by cement and construction activities thus becomes a way to envision a brighter future for the city and its surrounding areas. Boeri not only talks about the need to responsibly self-limit and of *auto-limitazione* as a "form of suspending action, of building, of occupying space"

(2011, 110; *forma di sospensione dell'agire, del costruire, dell'occupare*) in order to see such a future; he also borrows a word from the fields of biology and botany, graft (*innesto*), to define the sort of shared cohabitation between human beings and other animal species he has in mind, one in which all entities can healthily grow together. Thus as a way to conclude this section with a hint at its beginning, I wish to take the cue from this arboreal word and, while retrieving the image of Expo 2015's futuristic, bombastic, and high-tech sculpture of the *Tree of Life*, invite you to join me in thinking about a very different—though allegedly also more responsible—kind of sculpted treelike artwork.

Henrique Oliveira's astonishing *Transcorredor* (2016), also displayed at the *End of the World* exhibition in Prato, is a walkable, tunnellike, twisted structure made using a combination of mortar and mostly recycled materials (scraps of plywood, tree branches, bricks) that, depending on the way you look at it and enter it, resembles either a modern construction site (collapsing or incomplete) or an enormous, primitive-looking tree that has fallen to the ground. Walking around and through it may evoke a fantastic, utopia-like, nonchronological journey in space and time. It combines the organic with the inorganic, adopting assemblage and recycling rather than high-tech materials as its constitutional principle. Thus *Transcorredor* seems both to capture the new cosmopolitical horizon alluded to by Boeri and to be more in sync with our current environmental situation than the *Tree of Life*. It serves as a powerful reminder that nature and culture are part of an unfinished continuum, with a more "limited" and sustainable vision. And finally, it embodies the idea that what may seem like a scary descent into the underworld may also turn out to be a gradual journey from which we emerge perhaps with some hope for, or at least some useful insight into, the future.

Transcorredor's invitation to walk through and rethink the notions of construction, habitat, and mobility in environmentally friendly ways resonates with additional artistic gestures that encourage us to embrace a more restrained philosophy of incompleteness rather than one of feasibility that is just focused on increasing the monetary value of land through overbuilding. Both Wu Ming 2's "path" books, *Il sentiero degli dei* (The gods' path; 2010) and *Il sentiero luminoso*, for example, pay attention to issues of soil consumption and to the conflicts related to the perception and planning of living spaces. As Wu Ming 2 convincingly (2016) observes via Gerolamo (his autofictional,[22] wandering alter ego), it would be good to assign a positive rather than a negative connotation to the notion of neglect (*abbandono*) as it refers to territory, since it is often when and where a space is left to exist on its own—be it a muddy ditch on the side of a road rich with a surprising biodiversity or an urban area not yet colonized by some developer—that spontaneous alternative forms of spatial creativity and space utilization may be imagined and take shape.

As we shall see in the next section, however, both this sort of measured philosophical position regarding the use of land and any respect for a natural status quo are completely inconceivable, if not utterly blasphemous, to the protagonists in Alessandra Sarchi's *Violazione* (Violation; 2012), a novel set in the geologically unstable landscape of the Appenninic hills around Bologna that addresses, from yet another perspective, Italy's construction frenzy and omnipresent *horror vacui*.

Under the Landscape: Environmental Concerns in Alessandra Sarchi's *Violazione*

> Here all that's left is to wrap the landscape around the self and turn your back.
>
> Qui non resta che cingersi intorno il paesaggio qui volgere le spalle.
>
> —Andrea Zanzotto, "Ormai"

In his aptly titled first collection, *Dietro il paesaggio* (Behind the landscape; 2011a), and then repeatedly throughout his poetry and essays, Italian poet Andrea Zanzotto reflects on and captures the circular relationship that exists between landscape (which he imagines as a physical and spiritual entity) and human beings. He suggests that there are cognitive, psychological, and emotional links connecting individuals with places and communities with landscapes and that the latter constitute the existential horizon within which humans live, grow, and cultivate their self-awareness. In a dynamic that brings together the "earth" and the "flesh" (to refer to Franco Arminio's initial quotation), landscape is also "exposed to the danger of disintegration at times of deep crisis, be it social or individual" (Zanzotto, quoted in Giancotti 2013, 11; esposto al pericolo di una disintegrazione nei momenti di forte crisi, sia sociale che individuale). Zanzotto gives an anguished and concerned look at the advent of modernity and the economic boom in post–World War II Italy while striving to recuperate, at least at the linguistic level of his poetry, what he perceives as a primordial but also rapidly vanishing relationship among humankind, land, and nature.

As Patrick Barron puts it, Zanzotto often "delves beneath the surfaces of nature, history and language [and] explores the complex inter-weavings of culture and nature evident in his village [and] surrounding countryside" (2007, 3). Zanzotto's reach and message extend beyond the regional borders of his native Veneto, since his texts might well relate to other places where erosive and consumptive dynamics, not unlike those we have just witnessed in the Italian northwest, are at play. In this sense, one could say that a similar attempt to look at what is behind and, as my blatantly Zanzottian title indicates, beneath the facade of the *bel paese* characterizes also Alessandra Sarchi's poignant first novel, *Violazione*.[23] Another instance of a "postmodern engagement," the events in the novel signal the author's awareness of sociopolitical, ethical, and environmental questions.[24] Ideally following Zanzotto's lesson and participating in an imaginary dialogue on building and dwelling with the other authors considered in this chapter, Sarchi's text may be read as a meditation on the meaning of a specific place and on the repercussions of trying to dwell there at a time when the sort of communication mentioned previously—among humankind, land, and nature—is perilously close to being interrupted. As the novel uncovers what resides under the geological layers of earth, beneath the foundations of a newly built country house, and metaphorically, beneath the hypocrisy, complacencies, and behaviors of some of its main characters, it provides an apt illustration of ecosemiotician Timo Maran's words: "If people are illiterate or ignorant of 'the language of landscape' or misinterpret it, cultural, social and environmental failures often follow" (2010, 81).

Mostly set in the city of Bologna and the Apenninic countryside in the center-north region of Emilia-Romagna—one of the most educated, progressive, wealthy, and sociopolitically aware places in Italy, where "bicycling and trash recycling are encouraged" (Donati 2016, 169)—*Violazione* reminds us of the pervasiveness of illegal environmental practices and of greedy land developers crowding the Italian Peninsula. It also tells us that a disregard for ecological and ethical concerns is buried deep under the acquiescence, apathy, and self-serving behaviors of even those who (supposedly) should be far from "ignorant or illiterate" when reading and interpreting the "languages of nature" (Maran 2010, 85). Divided into three sections titled "Mondo" (World), "Casa" (Home), and "Frana" (Landslide), which allude to both a universal "primordial and allegorical condition" (Zinato 2012) and specific localized events, the novel recounts the interrelated vicissitudes of three families: the Donellis, an educated, progressive urban couple with two children eager to leave the city who find a new home in the countryside; the Draghis, building contractors and owners of the farm estate the Cinque Pini (Five Pines), where the Donellis' house is eventually built; and the Moldovan immigrant family of Natasha and her son, Jon, who work as the Draghis' house- and groundskeeper, respectively. As the book's title suggests, it tells a philosophical tale of the violation of natural and cultural laws, of building speculation and environmental abuse, of the violence and danger that are both naturally present in and cynically brought onto the territory, of our relationship with the nonhuman (land, animals) and the "other" (immigrants, the disabled), and of the unethical dimension of the space we inhabit.

Despite the centrality of human characters, *Violazione* also deals with the consequences of the encounter and the homology between human beings and nonhuman matter—be it artificial, like the concrete necessary to build the foundations and the herbicides used at the Five Pines to grow more vegetables, or natural, like the minerals that shape the landscape itself and the various animals and plants that, like epiphanies, mark crucial moments in the narration. As such, the novel encourages an approach informed by the conceptual framework of both material ecocriticism and ecosemiotics. If, according to the former, matter is perceived as a vibrant force possessing the capacity to effect change and thus achieves the status of an active agent in the dynamics of events, the main objective of ecosemiotic studies is to broaden the concept of text to include the structures of nature. As Timo Maran puts it, ecosemiotics' scope "can be expressed as a study of semiotic relations between an organism and its environment, various interpretations and representations of nature, communicative processes between human culture and living nature and problems in these, or culture's relations with the local environment" (2010, 83). In what follows, I contend that Sarchi's "moral novel," either knowingly or unknowingly, touches on these issues in a creative and thoughtful way, illustrating how such theories may apply to this additional Italian context.

Exemplifying a classical representation of nature's voice and agency and of its fraught relationship with human beings, the epigraph from Giacomo Leopardi's *The Moral Essays* (1983) immediately signals the novel's main topic and intentions.[25] *Violazione*'s initial pages expand on Leopardi's imaginary conversation as they suggest that the fundamental building blocks that constitute the human and the nonhuman, bodies and landscapes, are constituted by the same elements and materials. The first part of the text alludes thus to "a cosmic and non-anthropocentric habitat where matter, be it organic or inorganic, is at once matrix, mater and enemy" (Zinato 2012) and where there is no apparent gap between human and nature, landscape, and architect.

As Primo Draghi and Jon—his Moldovan antagonist and, in the end, victim—wake from their sleep, their bodies are still far from each other, separated by the mineral, lithic corporeality of the Carpathians, "a large mountain chain, that became the sore back from which many pains came to Europe" (Sarchi 2012, 13; una grande catena di monti, diventata la schiena dolente da cui venivano molti mali all'Europa). This association of the mountains' geological body to a biological (human) one via the word *schiena* (back) wonderfully captures an idea voiced by Manuel DeLanda: "Bone . . . never forgot its mineral origins: it is the living material that most easily petrifies, that most readily crosses the threshold back into the world of rocks" (1997, 27). As Primo and Jon slowly materialize on the page, they also emerge from an indistinct, preconscious, primordial, spatiotemporal dimension where everything is still related and entangled, where differences between living and nonliving entities depend on random combinations of proteins and substances or "a degree of stratifications" (261), and where identities and ontologies are still unfixed, fluid, hybrid entities under construction:

> A body of a man. . . . The sleep he tries to shake off is thick and murky like the aquatic broth in which molecules of proteins built the earliest forms of life over geological eras in an unstoppable chain of reactions . . . the man tries to lift a weight off himself . . . this is the weight of the millions of years the substances stirring in the primordial liquid took to aggregate and create complex systems, plants, animals, humans . . . he cannot recognize anything. He is a body in the body of the world. He could enjoy the blurring that tells him that he is a living being among all the other things that live but, instead, he is afraid. . . . So he exists, but who is he? Who is he?

> Un corpo d'uomo. . . . Il sonno da cui cerca di liberarsi è spesso e limaccioso come il brodo acquatico in cui, per effetto di un'azione inarrestabile e di conseguenti reazioni a catena, molecole di proteine formarono nel corso di ère geologiche le prime forme di vita . . . l'uomo prova a togliersi un peso . . . il peso dei milioni di anni impiegati da ciò che si agitava nel liquido primordiale per aggregarsi e dar vita a sistemi complessi, forme distinte, piante, animali, umani . . . non sa riconoscere niente. È un corpo nel corpo del mondo. Potrebbe godere dell'indistinzione che gli fa percepire di essere

vivente in mezzo al resto che vive, invece ha paura.... Dunque è, ma chi è? Chi è lui? (Sarchi 2012, 7–9)

Indeed, DeLanda's views of the nonlinear interactions between historical and natural processes and, in particular, his discussion of history "not in terms of 'man,' and his manifest destiny" but in terms of "coexistences of accumulated materials of diverse kinds, as well as by the processes of stratification and de-stratification that these interacting accumulations undergo" (1997, 265, 268) resonate well with these early passages in Sarchi's novel. While we begin to recognize the individual human and nonhuman forms that this "aquatic broth" (or, in DeLanda's words, this primordial "matter-energy"; 17) will take as it "awakens" and "intensifies" (265) over time, we also learn that part of this entangled stuff eventually decomposes and returns to the original geobiological mesh. The following description of Primo's dead father illustrates such a transition: "For longer than an hour he had begun to be a little less his father and a little more decomposing matter, similar to the farm animals they had slaughtered together" (Sarchi 2012, 25; Da più d'una ora aveva cominciato a essere un po' meno suo padre e un po' di piú materiale in decomposizione, non diversarmente dagli animali della fattoria che insieme avevano ucciso).

Of course, some of these matters can be defective and, as a consequence, unable to achieve their full potential. As a case in point, Vanessa, the younger of Primo's two daughters, emerged out of her mother's womb "as if made of waste materials" (Sarchi 2012, 10; sembrava uscita con materiali di scarto) and gives us an early, somber warning of the consequences of behaving irresponsibly, arrogantly, and self-centeredly when creating something—be this "thing" a new human being, a new building, or a new vegetable garden.[26] We soon find out as well that Primo's fear about his loss of identity ("he feels shapeless, like the animals that already lived on Earth" [9; si sente informe come gli animali che già vivevano sulla Terra]) is not shared by Jon, who surfaces from the depths of his own sleep in his uncle's apartment in Bucharest. Rather than being anguished because of his undefined situation (or perceiving it as an oppressive weight), Jon feels "the gratitude of refugees" (13; gratitudine dei rifugiati). Even though, like Primo, he momentarily loses his identity, Jon retains the fundamental awareness that despite his spatiotemporal coordinates, his condition will always be that of a guest (*ospite*) in the world. This attitude allows Jon to simply accept both his vulnerable existence and—on the same horizontal, ontological level—the existence of all the nonhuman things that surround him:

> I did not know who I was, I existed and that's it, in the same way as this room, this blanket exists....
>
> He began to ... observe the objects that filled the room.... Suddenly the faded wallpaper seemed to be good and, similarly, so did the opaque windowpanes, the stereo.... Everything that surrounded him was pervaded by a primordial kindness. Or perhaps these things seemed kind to him because they were helpless, because he saw himself as helpless.
>
> Non sapevo chi ero, esistevo e basta come esiste questa stanza, questa coperta....
>
> Cominciò a ... osservare gli oggetti che riempivano la stanza.... Di colpo la tappezzeria stinta gli sembrava buona, i vetri opachi della finestra gli sembravano buoni pure quelli, buono l'impianto stereo.... Tutto ciò che lo circondava era pervaso da una bontà primordiale. O forse le cose gli apparivano buone perché le vedeva inermi, perché inerme vedeva se stesso. (14–16)

Thus a crucial and far-reaching opposition is immediately created in the novel, as the appearances of these two characters point to radically antipodal ways of perceiving one's place and being-in-the-world. On the one hand, Primo epitomizes a traditional, thoroughly anthropocentric position founded on "the sovereignty of the [knowing] subject ... the transparency of language; and rationalism" (Halliwell and Mousley 2003, 4). On the other hand, Jon embodies a view of himself and of "humans, animals, and things as ontologically bound up in reticular arrangements with similar and not so similar forms" (Watts 2013, 3).

From an ecosemiotic perspective—namely, one that wishes to "open the concept of text to natural phenomena" (Maran 2010, 81)—these two characters possess very different hermeneutical skills and degrees of literacy in terms of reading the environment they inhabit. Primo's incapacity to see, feel, and recognize his surroundings as he strives to understand who and where he is ("He touches the sheet with his hand, but it is a blind touch.... He does not see and does not hear ... he is not able to recognize anything" [Sarchi 2012, 7–8; Tocca con la mano il lenzuolo, ma è un toccare cieco ... non vede e non sente ... non sa riconoscere niente]) is juxtaposed with Jon's progressive perception, interpretation, and nonhierarchical valorization of a variety of objects, both nonhuman and human. As sights and sounds interrelate with one another and with Jon, they ultimately contribute to shaping his own identity. To his uncle's question "How did you remember who you were?" (Come hai fatto a ricordarti chi eri?), Jon answers,

The table made me. . . . Then my mother . . . her image, I don't know how, she gave me my name back . . . he felt the sweetness of names, persons and things, which one needs only to pronounce to be alive.

È stato il tavolo. . . . Poi mia madre . . . la sua immagine, non so come mi ha restituito il nome . . . sentí tutta la dolcezza dei nomi, delle persone e delle cose, che basta pronunciare perché siano vive. (14)

His words illustrate the distinction made previously, especially if one underlines their egalitarianism rather than their apparent belief in the reality-giving power of language.[27]

However, Sarchi soon makes clear that Primo's ignorance is by choice rather than by nature, and therefore, it is ethically charged. For example, as Primo drives back from the hospital after seeing his father's corpse, he too senses that there is a continuity between body and land, the human and the nonhuman; similarly, he cannot escape the display of environmental plagues that affect Bologna's suburban area. Nevertheless, he ultimately opts for a disturbing, selective blindness:

The skin on his face was dry and wrinkled, stretched on two pointy cheekbones that looked like poles stuck in the dirt . . . his feet . . . were absurdly pointing up and reminded of roots torn off the ground. . . . The meters of asphalt in front of him looked like obstructed veins, yielding, unable to hold, and similar to his father's black arteries. It's spontaneous to turn one's head elsewhere. . . . [There are] enormous commercial centers anchored to the ground as if they were scarabs with their armor shining under the sun. Trailer parks, rental cars parking, curtains hung to dry and covered with fine dust, carbon monoxide, lead, benzene, in short, the sorts of combusted hydrocarbons that make the air yellowish and gray . . . plastic bags and dishes on the roadside, cigarette boxes, bottle and cans at the bottom of a ditch. . . . At night, of course, at night is different. Primo likes to drive at night.

La pelle della faccia arida e rugosa, tenuta su da due zigomi appuntiti come pali conficcati nel terreno . . . i piedi . . . assurdamente rivolti in alto facevano pensare a radici divelte. . . . I metri di asfalto che ha davanti assomigliano a certe vene ostruite che cedono e non tengono, come le arterie nere di suo padre. Viene spontaneo girare la testa altrove. . . . Enormi centri commerciali ancorati al suolo come scarabei dalla corazza lucida sotto il sole. Depositi di caravan, autonoleggi, tende stese ad asciugare e impregnarsi di polveri sottili, monossido di carbonio, piombo, benzene, insomma tutta la varietà degli indrocarburi combusti, che danno quella patina giallognola e grigia all'aria ... i sacchetti e i piatti di plastica sul bordo, le scatole di sigarette, le bottiglie e le lattine sul fondo del fosso. ... Di notte, certo, di notte è un'altra cosa, a Primo piace viaggiare di notte. (Sarchi 2012, 17–19)

This cognitive attitude is not surprising in a rapacious, self-centered character like Primo, who grew up in a backward Southern peasant family who had emigrated North and made their fortune through construction and speculation and whose idea of progress includes the wider availability of dichlorodiphenyltrichloroethane (DDT) and other chemicals to kill bugs and weeds in their newly purchased estate farm:

> DDT and all the insecticides for the house had made Primo's mother glad after the war when ... she had moved from Molise to Emilia. ... At the local farmers' co-op, there were liquids and powders to spray that got rid of the nuisance and the damage caused by all the insects hidden amidst the vegetables and the weeds.
>
> Il Ddt e tutti gli insetticidi per la casa avevano fatto la sua (i.e., Primo's mother) gioia nel dopoguerra quando ... si era trasferita dal Molise all'Emilia. ... Qui al consorzio, esistevano liquidi e polveri da spruzzare che eliminavano il fastidio e il danno di tutte le bestie annidate in mezzo alla verdura e le erbacce.[28] (42–43)

This intentional shortsightedness and the relative ease with which people turn their heads to the environmental consequences of their actions are less expected and more problematic when they are unearthed in the cultivated and sociopolitically aware left-wing family of Linda (a neurologist) and Alberto Donelli (an employee of the region's environmental agency). The couple's book collection includes Descartes, Yourcenar, Bergson, Volponi, and even one of the forefathers of biosemiotics, Jakob von Uexküll. This advanced literacy, however—which manifests in their interpretative skills, ecological concerns, and green-living aspirations—is slowly buried by a familiar, self-serving tendency to compromise and abuse their reason in order to justify desirable but also morally questionable decisions.[29] Implicitly reaffirming the relationship between the body and land and the human and nonhuman, Sarchi skillfully tracks the couple's intentional visual and cognitive obscuration while bringing to light the secrets concealed under the mix of cement, "limestone, clay and chalk" (2012, 59; calcare, argilla e gesso) at the Five Pines.

Environment and Behavior, the title of von Uexküll's book in Linda's library, could well describe the central and final sections of the novel. These map the Donellis' eventual move from the city to the countryside and their parallel descent into literal and cognitive darkness. At the same time, these pages also illustrate once more the "multiple coexistences and interactions" between the natural and the constructed, the inside and the outside—between "geologic, organic and linguistic materials" (DeLanda 1997, 21). Thus as we learn revealing details about Linda's past and present life and abodes—her nostalgia for a fig tree now reduced to a name on a street sign (via del Fico), its roots buried "like the selenite was buried in the cellars" (Sarchi 2012, 59; come la pietra di luna era sepolta nelle cantine); her rationalist, anthropocentric stance ("At the top of everything was the human brain" [62; In cima a tutto stava il cervello umano]); her fashionable, self-centered rehabilitation of "the vegetable world in the hierarchy of being" (63; il mondo vegetale nella scala dell'essere)—we also better realize that nature here is "that which lays beneath, above and inside ourselves" (Pugno 2012, 11; ciò che sta sotto, sopra e dentro di noi). Sarchi writes,

> Linda was born and grew up among solid red bricks . . . a system of vaults that supported the building and pieces of wall with slabs of selenite had emerged from under the floor during some restructuring work . . . she preferred to call moonstone the glimmering selenite. It was the same stone one could see in the foundations of towers, in the city walls and now it came up also in their cellar . . . it had arrived there on the back of donkeys, down from the hill where it surfaced in abundance amongst the gullies. . . . After getting married they went to live . . . in the heart of the city, where selenite, sandstone, red brick and finally concrete had covered up the ground, shaped by the most ingenious thoughts, in turn ingeniously hiding their destructive potential.

> Linda era nata e cresciuta tra solidi mattoni rossi . . . erano emersi sotto il pavimento un ulteriore sistema di volte che sorreggeva l'edificio e tratti di muro a secco con qualche lastra di selenite . . . quella pietra luccicante, la selenite, che lei preferiva chiamare pietra di luna. Era la stessa pietra che si vedeva nei basamenti delle torri, nelle mura di cinta della città e ora saltava fuori perfino nella loro cantina . . . c'era arrivata a dorso di mulo, trasportata giú dalla collina, dove affiorava in abbondanza in mezzo ai calanchi. . . . Dopo essersi sposati erano andati a vivere . . . nel cuore della città, dove

la selenite, l'arenaria, il mattone rosso e da ultimo il cemento avevano ricoperto la terra, prestando la forma ai pensieri più ingegnosi, occultando in maniera altrettanto ingegnosa il loro potenziale distruttivo. (2012, 58, 62)

This passage clearly points to the difficulty of separating the natural from the cultural and, in turn, to the dangers of disregarding—or pretending to conceal under an artificial patina—the timeless, vibrant energy of matter. In Linda's childhood home, sandstone, bricks, and concrete are only apparently under the control of human imagination, since they retain a destructive potential that points to their inherent agentic force.

Linda's romanticized perception of nature and subsequent eagerness to check out the country house is not shared by her husband, Alberto. Paradoxically, the frustrations and compromises linked to his job, combined with the realization that the countryside, city, and mountains are now parts of an extended nonplace,[30] make him think that "ecology is a lost battle. Green is an illusion . . . nature . . . a great abstraction" (Sarchi 2012, 71–72; l'ecologia è una battaglia persa. Il verde è un'illusione . . . la natura . . . una grande astrazione)—a lost Arcadia identifiable only in TV commercials. And yet he still wants to believe that "the country was still wonderful, and part of such wonder should be theirs" (69; il paese fosse ancora meraviglioso, e che una parte di quella meraviglia dovesse toccare a loro). After all, Alberto's own, most intimate nature—similar to the dirt of the hillside gullies (*calanchi*) that are characteristic of the rural landscape just outside Bologna—is literally to be pliable, to cave in, to yield, to subside.[31] The following passage, in which he notices blocks of glimmering selenite and quartz crystals along the road to the estate and reflects on the fact that the same carbon-based material is also a crucial component in the human brain, captures this overlap between Alberto and the landscape and underlines the transcorporeal entanglement between the human and the nonhuman, the organic and the inorganic:

> Alberto is tempted to stop the car, get down and observe more closely the sequences of $CaSO_2H_2O$. . . . It is the same matter: our brain is an accumulation of carbon mixed with oxygen, soft and spongy rather than polished and hard like quartz.
>
> Alberto è tentato di fermare la macchina, scendere e osservare da vicino le sequenze di CaSO2H2O. . . . Si tratta della stessa materia: il cervello è un accumulo di carbonio

attraversato da ossigeno, molle e spugnoso anziché lucido e resistente come il quarzo. (98)

This passage also tells us that Alberto has excellent observational skills and that, to a greater extent than his scientist wife, he is quite literate in interpreting the "language of landscape." He displays the same kind of visual attention during a visit to their soon-to-be new house on the Draghis' estate. On this occasion, however, a series of warning signs are systematically ignored so that the Donellis' domestic green dream might be realized. For example, when Alberto sees numerous cans of dangerous chemical pesticides in a place that should be a certified organic farm, he "decides not to care and resolutely turns his head away" (Sarchi 2012, 117; decide di lasciar perdere e si volta in maniera decisa da un'altra parte), behaving just like Primo does earlier when he covers up the unsightly view of trash along the road. Similarly, when Linda, the neurologist, wonders about the origin of Vanessa's disability, she only chooses to see the educational shortcomings of the girl's mother (Genny) without acknowledging her own shortsightedness and unwillingness to admit that an environmental cause could also be responsible for the girl's condition. A tour of the farm's large stable and its animals, followed by the sight of other animals "slaughtered and frozen" (133; fatti a pezzi e congelati) in an industrial refrigerator, and then offered to the Donellis in the shape of food ("ground beef, chops ... and steaks" [133; macinato, costine, carne da brodo e bistecche]) represents another significant warning sign for Linda and Alberto. However, although this vision is a "punch into their stomach" (133–34; pugno nello stomaco), it is not enough for them to realize that the distance that separates the violence toward animals, the landscape, and humans is very small.

A perfect emblem of human behaviors and widespread attitudes in the Anthropocene, the Donellis' inherent predisposition to know and see and yet, at the same time, to leave themselves open to misinterpretation, self-deception, and indifference finds its culmination when Alberto discovers that the stable was built over a filled-in river in a dangerous, hydrogeologically restricted area. Linda's response to this piece of news ("In the end they only built a stable; it doesn't look that bad" [Sarchi 2012, 170; In fondo hanno solo costruito una stalla, non mi sembra cosí grave]) reveals at once both the twisted, material wilderness that characterizes the landscape of the Five Pines and the immaterial, rhetorical concealment of any remnant of legality and morality.

As the novel moves toward its catastrophic ending, a flashback informs us that the stable is not the only illicit construction on the estate:

> Urbanize and sell.... The first leg in this project had been the restructuring of the ruin in the middle of the valley.... On a Sunday morning [Primo] had gone to the valley with one of his powerful machines ... it did not take long to complete the dismantling and to pile up on a side the debris that would be used as foundation material for the new house ... he decided to plane also the strip of land that always seemed ready to fall onto the ruin. This was a knoll that worked as a sort of cushion between the ruin and the back of the larger hill.... Only the stupidity and misery of the peasants ... could have convinced them that such a useless and ugly embankment was needed ... [Primo] had dealt more and more powerful blows until he eroded the foot of the mountain and, underneath, found layers of gravel and [also] that the ancient flood bed of the river Reno extended up to that area.... The future house had gained at least ten meters on the back ... more money that would fill his pockets.

> Urbanizzare e vendere.... La prima tappa di questo progetto era stata il rifacimento del rudere in mezzo alla valle.... Una domenica mattina [Primo] era sceso nella valle con una delle sue potenti macchine ... non gli ci volle molto per completare l'opera di abbattimento e ammucchiare su un lato i detriti che sarebbero serviti per fare la fondazione della nuova casa ... decise di piallare anche la lingua di terra che sembrava sempre in procinto di rovesciarsi sul rudere. Era una montagnola che faceva come da cuscino tra il rudere e la schiena della collina vera e propria.... Solo la stupidità e la miseria dei contadini ... potevano averli convinti a mettere un terrapieno inutile e brutto ... [Primo] aveva assestato colpi sempre piú aggressivi fino a erodere il piede della montagna, scoprendo che sotto c'erano depositi di ghiaia e che l'antica area golenale del Reno doveva estendersi fino a lí.... La futura casa aveva acquistato almeno dieci metri sul retro ... altro denaro che sarebbe entrato nelle sue tasche. (Sarchi 2012, 180–82)

Primo Draghi's main mission in life (to build and sell) provides him with a very limited understanding of his surroundings—one that disregards the land's history, agency, memory, and culture. In this sense, it remains at the opposite end of the *umwelt*-influenced notion of "private landscape" put forth by Almo Farina. With this expression, Farina indicates a subject-centered perspective that can take "into account the multiplicity of agencies ... in a living environment" and aims at reducing "the gap between human values and ecological processes" (quoted in Lindström, Palang, and Kull 2013, 104). In Sarchi's novel, rather, the landscape is thoroughly privatized, and as such, ecology (from the Greek *oikos*, "house") is subordinated to the interests of the economy (i.e., the house's management and *nomos*, "law").[32] Within this context, when the landscape responds to the aggressions it has been suffering (a landslide that results in a hole in the ground close to the planned building site), Primo's first reaction is to conceal the problem by covering the site with more dirt and building over it. He knows from experience that once the earth is silenced and the sales contract is signed, the landslide and its potential dangers will no longer be his problem: "He had replied to the violence of the hill with equal violence, silencing it, regimenting it with poles stuck deep in the ground" (Sarchi 2012, 220; Aveva risposto alla violenza della collina con la stessa violenza, mettendola a tacere, irregimentandola con pali conficcati in profondità nel terreno). As Primo remarked earlier, "The landslide will be the Donellis' problem" (186; La frana sarà un problema dei Donelli).

As it turns out, after the Donellis finalize the purchase and move into their new place, the landslide remains an invisible, underground event within their new, superficial (in every sense of the word), existential horizon. Only Jon, who in the meantime has reunited with his mother and works for Primo as a stable boy, is able to see this hidden wound in the land:

> The vegetation had covered and perfectly camouflaged that week of frantic work during which they had excavated and then poured in concrete and iron in order to plug the hole that the earth had dug underground. He was confident that the family who had bought the house didn't know anything.
>
> Il verde aveva rivestito e camuffato alla perfezione quella settimana di lavoro forsennato in cui avevano sbancato e riportato cemento e ferro per tappare il buco che la terra aveva scavato sotto di sé. La famiglia che aveva comprato la casa non sapeva niente, ne era certo. (Sarchi 2012, 229)

As someone who thinks of himself as just one more vulnerable presence in the world, on par with other living and nonliving things, Jon's ability to read the landscape as well as his ease with and understanding of the animal realm is correlated with his status as an immigrant and resident alien. On the one hand, he is just another young Eastern European trying to make his living at the margins of Italian society, invisible to the law; on the other, he is an emblematic crosser of boundaries, frontiers, and identities, both physical and metaphorical. In this sense, some of Rosi Braidotti's observations on the notion of "nomadic subject" seem applicable here. As she reminds us, "The polyglot is a linguistic nomad. The polyglot is a specialist of the treacherous nature of language, of any language.... Nomadism consists not so much in being homeless, as in being capable of recreating your home everywhere" (1994, 8–17).

Despite my previous observations regarding Jon's mostly sonic relationship with language, both nomadism and multilingualism are characteristics that fit him well. He not only quickly adapts to his new living conditions and befriends the Donellis' children, thus re-creating an ephemeral home in his new surroundings, but he is also a specialist in understanding (besides Italian and Romanian) the nonhuman language of animals and the signs of nature. Although it would be excessive to affirm that he literally merges human and nonhuman aspects, he certainly displays the clearest features of a natural-cultural blend. After all, he is equally literate whether he is "under a tree with a book in his hands" (Sarchi 2012, 77; con un libro in mano sotto un albero), dealing with horses, deciphering the droppings of wolves on the property, reading the markings of deer on a low tree branch, or communicating with the Draghis' dog. In short, Jon appears to be one with both culture and nature, and there is no definite border between him and his surrounding environment.

Jon's final act of standing up to Primo, his refusal to acquiesce to another one of the Draghis' illegal building enterprises, distinguishes him as the only wholesome and ethically sound character in the novel. Unsurprisingly, this gesture, combined with Jon's awareness of Primo's covering up of the landslide, is also his death sentence, since Primo, enraged, eventually shoots him. As the same kind of illiterate and destructive violence Primo displays toward the land shifts toward Jon, the latter's dead body becomes the clearest material instance of the social and environmental failures occurring in this place. As Jon's desperate mother, Natasha, intuitively understands, Primo is just the most visible cause of her son's death. The violated land itself also has some underground role in this tragedy: "Natasha ... turned around with wide-open eyes as if her son's killer were everywhere, in every clod of that earth" (Sarchi 2012, 260; Natasha ... si girò intorno con gli occhi sbarrati come se l'assassino di suo figlio fosse ovunque, in ogni pezzo di quella terra). As the killer's identity is scattered over the wounded earth, and Primo is implicated in Jon's death, the tragic finale of the novel thus evokes once more the blurring of the human and the nonhuman, earth and flesh.

In the end, *Violazione* tells us that when human beings make a habit of disregarding ethical values and inflict violence on the land, sooner or later, a scarred environment will in turn respond with violence without making distinctions between who is guilty or innocent, powerful or vulnerable, good or evil. Eventually, everybody will pay a price, since ultimately, everyone and everything—"A river, a tree, a kid less than twenty years old" (Sarchi 2012, 267; Un fiume, un albero, un ragazzo di nemmeno vent'anni)—actually *is* landscape and linked to the well-being or ill-being of the ecosystem.[33] As Salvatore Settis observes in *Paesaggio, costituzione, cemento* (Landscape, constitution, cement), a new space "dominated by money and the market kills everyone's story, violates everybody's life" (2010, 55; dominato dal denaro e dal mercato uccide la storia di tutti, violenta la vita di ognuno). This remark could not be more appropriate to describe the conceptual stance underlying Sarchi's novel.

The novel's tragic conclusion, however, may also offer us an occasion to reassess values and meanings and, expanding our horizon from the local to the global, to think of it as if it were a sort of apocalyptic event enabling us to disclose something that was not fully comprehended before—to (re)discover and dig out a lost sense of measure (or both sense and measure) from the chaos.[34] Now experiencing a sensation of destabilizing *unheimlichkeit* (uncanniness), far from the one he and his wife expected to have in their new country home, we learn that Alberto Donelli is eventually able to transcend the localism of his own situation and thus "does not only think of cohabiting with humans, but also with the earth. There were many telling signs about a loss of measure. The human animal had to be redomesticated" (Sarchi 2012, 270; non pensa solo alla convivenza con gli uomini, ma anche a quella con la terra, c'erano tanti segni che dicevano che s'era persa la misura. L'animale uomo andava riaddomesticato).

Although we are not told how this attempt to restore measure, to refamiliarize ourselves with the interpretation of signs, and to redomesticate ourselves should proceed, the same fiction to which Alberto belongs constitutes a first step in this crucial process of recivilization. Serenella Iovino writes that "to imagine a place is the first step to re-inhabiting it" (2012b, 104). Imagining a different way of cohabiting with our planet does seem to be a good idea if, learning from Sarchi's compelling novel and having lost what Ernesto De Martino calls the "domesticity of the world" (2002, 479), we wish to avoid finding ourselves in a literal and metaphorical state of homelessness while being surrounded by concrete.

186 Ecologies of Mobility in Contemporary Italian Literature: Simona Baldanzi's *Mugello sottosopra* and Mauro F. Minervino's *Statale 18*

> Since the time of my childhood, I haven't spent a day in Mugello without seeing a construction site. Always more insistently, I have seen the gray of concrete mixing with the green.
>
> Da quando sono piccola, non ho passato un giorno in Mugello senza un cantiere. Ho visto il grigio del cemento mescolarsi al verde con sempre più insistenza.
>
> —Simona Baldanzi, *Mugello sottosopra: Tute arancioni nei cantieri delle grandi opere*
>
> One thing is clear around here. That cement lays at the center of every interest.
>
> Una cosa è chiara da queste parti. Che al centro di ogni interesse c'è il cemento.
>
> —Mauro F. Minervino, *Statale 18*

In the years after World War II, then during the "economic miracle" in the sixties until today, Italy's construction frenzy, love for bricks, and eagerness to modernize by indiscriminately covering thousands of square kilometers with asphalt and cement—Cederna calls it "roadmania" (quoted in Turri 1979, 129; stradomania)—affected some regions more than others. In 1979, Eugenio Turri could actually observe, thus leaving room for some hope, that

> in Italy there have not been "miracles" everywhere, its landscape has not been transformed, renewed or offended everywhere by new objects and insertions. There is also an Italy that has remained immobile. . . . From Lazio to Calabria, to Sicily and Sardinia . . . it is the internal, mountainous part, the one left out of the tumultuous growth of the coasts.
>
> non tutta l'Italia è stata "miracolata," non dappertutto il suo paesaggio è stato trasformato, rinnovato od offeso da nuovi oggetti e inserimenti. C'è anche un'Italia rimasta immobile. . . . Dal Lazio alla Calabria, alla Sicilia e alla Sardegna . . . è la parte interna, montuosa, esclusa dal tumulto delle coste in rapido aumento. (237)

Were he writing today, Turri would need to substantially revise his remarks, since the concerning level of soil consumption has spread beyond the Italian northeast into many other regions in the peninsula.[35] In these intervening years, in fact, the sought-after quiet and tranquility of several once "non-miraculated" areas has been displaced by a frenzy of construction activities and the desire to substitute what geographer Tim Cresswell calls a "sedentarist metaphysics" (2006, 26)—often associated with primitive, unsophisticated, and old-fashioned attitudes—with a modern one that is infused with a neoliberal rhetoric of so-called sustainable development in which mobility and "indiscriminate overbuilding" (Settis 2010, 54; indiscriminata cementificazione) play central roles.

Two areas of the Italian Peninsula are the focus of this final section. I draw my inspiration from a recent issue of the journal *ISLE* that is dedicated to the "ecologies of mobility," which considers how various modes of mobility may "devalue, distort, or destroy the natural world" or, alternatively, "'truly and livingly' encounter and know that world" (Withers 2017, 73). At the same time, I am inspired by Tim Cresswell's observations in his *On the Move: Mobility in the Modern Western World* (2006), particularly his idea that the study of mobility should go hand in hand with the analysis of the politics and materialities attached to movement.

The first area is the Mugello—the historic, lower Apenninic territory north of Florence, bordering Emilia-Romagna[36]—as it emerges in Simona Baldanzi's *Mugello sottosopra: Tute arancioni nei cantieri delle grandi opere* (Mugello upside down: Orange overalls in the building sites of the great works; 2011), a sociological inquiry that mixes narrative, pictures, graphs, interviews, data from official assessments, and newspaper articles in order to shed light on the difficult lives of the construction workers who built the railroad for the Treno Alta Velocità (TAV; a high-speed train) and the Variante Autostradale di Valico (VAV; A1 Highway Variant).[37] The second area is the Southern region of Calabria in cultural anthropologist Mauro F. Minervino's engaging "road narrative" *Statale 18* (State road 18; 2010), an ethnographic reflection on the material and cultural impact of one of the busiest roads in the Italian South.

Despite their geographical distance and cultural differences, these two places, Mugello and Calabria, immediately become closer if, with Baldanzi, we observe that the majority of the bodies working in Mugello's TAV and VAV building sites actually belong to male Calabrian migrants.[38] Additionally, both territories have been transformed and, as Turri would say, similarly "offended" by the construction of large transportation infrastructures and floods of concrete, bricks, and asphalt. These substances may be thus viewed both as agents of environmental degradation that have submerged preexisting landscapes and as examples of those "materialities attached to movement" mentioned by Cresswell (2006).

In Wu Ming 2's *Il sentiero degli dei*, as Gerolamo approaches the end of his cross-Apenninic walk from Bologna to Florence, he summarizes what he had earlier defined as the many "stomps that, in the past fifteen years, have crushed Mugello" (2010, 109; pedate che hanno schiacciato il Mugello negli ultimi quindici anni) and reflects on the human and nonhuman victims of the TAV, which now cuts through this previously uncontaminated area:

> Iron and concrete become the symbol of a negative utopia . . . of a territory without a landscape. . . . Even before departing, the supertrain has already made many victims: deaths on the jobsite, dried-up streams, vanished springs, wasted money. All these are crimes that refer to a bigger crime: the attempted murder of a difference. Namely, the one that distinguishes a place from a pile of dirt, a river from the trail of a motorway, a public road from a private villa's entrance, a man from his workforce.

> Ferro e cemento diventano il simbolo di un'utopia negativa . . . di un territorio senza paesaggio. . . . Il supertreno ha già fatto molte vittime, ancora prima di partire. Morti ammazzati sul lavoro, torrenti a secco, sorgenti sparite, soldi bruciati. Tutti crimini che rimandano a un crimine piú vasto: il tentato assassinio di una differenza. Quella che distingue un luogo da un ammasso di terra, un fiume da una sede autostradale, una strada di tutti da un ingresso di una villa, un uomo dalla sua forza lavoro. (145–46)

A similar kind of reflection lies at the root of and informs Baldanzi's investigation. Mirroring the concerns of the No TAV movement in Piedmont's Susa Valley, her text begins by listing the TAV's environmental damages to her native Mugello and noting the fundamental lack of difference in terms of visibility and risk exposure between the endangered mountain slopes and aquifers affected by the tunnels' construction and the blue-collar workers hired to dig them. As she puts it, "In the tunnels [the workers] remained invisible just like the aquifers and, perhaps, they too were in danger because nobody was considering them" (2011, 26; Nelle gallerie [i lavoratori] rimanevano invisibili proprio come le falde e, forse, proprio perché nessuno li prendeva in considerazione, erano a rischio anche loro). Indeed, as concrete pours over the Mugello, hiding human and natural bodies and erasing the differences between them under the tunnels' grayish vaults, I cannot help but think of Kara Thompson's definition of a sidewalk as "a concrete blanket that covers layers of soil, sediment, rock, and roots" and her related observation about the inevitable reappearance of what lies beneath, "whether we want it or not" (2019, 34). Be it in the form of water dripping from the tunnels' ceiling or revelatory narratives such as those considered here, some cracks (or "folds" in the blanket) do come to the surface and ask to be examined. Thus Baldanzi's main objectives are to consider the risks and degradation brought by the TAV and VAV to the territory, to the workers in the building sites, and to the local residents and to investigate the dynamics of the whole ecosystem, both above- and underground. By addressing the intrinsic links between environmental and social justice issues, *Mugello sottosopra* represents a recent example of Italian environmental justice writing.[39]

In this sense, the chapters dedicated to the workers' job-related diseases, injuries, and deaths are particularly relevant. They correlate the ecological damages—and the by now judicially established absence of environmental protection in the Mugello territory during the construction process—with a parallel lack of attention and protection for the men in the building sites, thus emphasizing the material interplay of their bodies with the surrounding places and substances. Baldanzi individuates many pathologies and "bio-psycho-physical deficiencies" (2011, 176; carenze di natura bio-psico-fisica) among the "orange overalls" that point to a degradation and vulnerability that are, once again, simultaneously biological, ecological, and socioeconomic. She suggests, for instance, that the workers' geographic provenance is closely related to the hazards and arduousness of their tasks and, in turn, to the severity of their diseases. Southerners tend to have the most burdensome jobs (miners, carpenters, etc.), while the workers from Tuscany are mostly "white collar or employed in complementary services" (172; personale quasi del tutto impiegatizio o legato a servizi complementari). She mentions "chronic fatigue . . . backache . . . hypertension, gastritis . . . depression" (173–74; fatica cronica . . . mal di schiena . . . ipertensione arteriosa, gastriti . . . depressione), feelings of isolation, and substance abuse and gambling addictions. In short, she describes a human corporeal landscape that is as degraded, exposed, and metaphorically excavated as the physical, geographical, nonhuman one.

Although materials like concrete and asphalt initially seem less likely to "infiltrate human bodies" (Alaimo 2010, 28) than, say, dioxin or asbestos, toxicity is still very much a part of this discourse. In light of the unlawful disposal of construction waste that contaminated Mugello's water and soil; the dusty, hydrocarbonated air in the tunnels; and the widespread abuse of alcohol and tobacco, what is especially toxic here is what Iovino calls "the pollution in the use of power" that is coterminous with any discourse on environmental justice in Italy (2016a, 6). A 2012 report by the Italian environmental association Legambiente entitled *Cemento S.p.A.* observes not only how the deluge of and pressure exerted by cement are responsible for the loss of almost half a square mile of soil every day but also how concrete's supply chain (*filiera del calcestruzzo*) is corrupt and entangled with mafia activities (Legambiente 2012). In this specific Tuscan context, we might then say that this sort of pollution manifests through the presence of discrimination, prejudices, and divisions based on one's regional origin; through a disregard for the workers' and the land's well-being; and unsurprisingly, as the trials and lawsuits Baldanzi mentions demonstrate, through criminal connivances between some of the construction companies involved and the political sphere.

At the end of the first part of her book, after focusing mostly on the tunnel workers, Baldanzi summarizes the TAV's impact on and alleged usefulness for the local inhabitants of Mugello by recounting a personal travel experience from her hometown of Barberino to Milan:

> I went by bus . . . in three hours it took me to the center of Milan. It was not convenient for me to take the "Frecciarossa" [the high-speed train] either to save on the cost . . . or on the time: [in order to take the Frecciarossa,] I would have needed first to take a bus to Florence . . . there is no railroad in Barberino. . . . After a story lasting fifteen years, the irreversible damages to the territory, the difficult life of the workers in the building sites, what should a local resident who takes a bus to Milan even after the TAV has been built think?

> Ho preso un autobus . . . mi ha portato nel centro di Milano in tre ore. Non mi conveniva prendere il Frecciarossa né per un risparmio economico . . . né per un risparmio di tempo: sarei dovuta scendere a Firenze con un autobus . . . perché a Barberino non c'è la ferrovia. . . . Dopo una storia durata quindici anni, i danni irreversibili al territorio, la vita difficile dei lavoratori dei cantieri, un abitante del Mugello che anche dopo la realizzazione del Tav prende un autobus per andare a Milano cosa dovrebbe pensare? (2011, 142)

Her rhetorical question here is useful to stimulate some conclusive considerations. What it reveals is not only the illogicality of the whole project from the perspective of a local traveler and customer but also the disjunction between the promise represented by the construction of a high-impact transportation infrastructure—which, according to its proponents, was supposed to improve connections between places, provide the "freedom to move" (Cresswell 2006, 15), and serve the mobility needs of all people—and its actual, much more limited (and limiting) function, since not everyone has equal access to it.

Of course, I realize it would be incorrect to assume that only a metropolitan "kinetic elite" is taking advantage of the TAV. However, Baldanzi's words suggest that there is indeed "a link between the mobility of some and the immobility of others" (Cresswell 2006, 255) or, to put it differently, that "transportation—due to its complicity in racism and other forms of institutionalized prejudice and inequality—only liberates *some* people" (Withers 2017, 68). Such "others"—here represented by the indigenous Mugello residents and the nomadic Southern migrant workers alike—paradoxically experience the same transportation challenges they faced at an earlier, pre-TAV time. Because of the TAV, they actually feel more disconnected, isolated, marginalized, and abandoned, and they often still need to take alternative, preexisting routes in order to go from one place to another.

From this perspective, the section narrating Baldanzi's own journey from Tuscany to Calabria, which she undertakes in order to learn more about the migrant workers and their commuting habits to Mugello, provides a striking counterpoint to the TAV's and VAV's repeated emphasis on technology and effortless, luxurious, air-conditioned speed. In particular, she refers to the A3 Salerno-Reggio Calabria, a southern section of the A1 Highway, whose status of perennial incompleteness and disrepair has reached quasi-mythical proportions among Italian drivers:

> Our journey would be a 14-hour nightmare of heat, thirst, lines, sweat, waiting, fatigue. We think of the Calabria that continues to emigrate, of the Salerno-Reggio Calabria as an enormous treadmill that watches them go by.
>
> Sarà un incubo di 14 ore di caldo, sete, code, sudore, attesa, stanchezza. Ripensiamo alla Calabria che continua a emigrare, alla Salerno-Reggio Calabria come un gigantesco tapis roulant rotto che li vede passare. (2011, 121)

As her words ultimately juxtapose two very different kinds of mobilities—one allegedly constituted by mostly urban travelers "who inhabit the luxurious space of flows" (Cresswell 2006, 256) and the other by emigrants on a run-down treadmill—the maxim "There are no TAV travelers without Calabrian emigrants" (an adapted version of Zygmunt Bauman's "There are no *tourists* without *vagabonds*," quoted in Cresswell 2006, 256) might aptly summarize this situation.

Expectedly, Baldanzi's general view on mobility when she discusses either the hardships faced by certain travelers—migrant workers forced to leave their homes and finding inhospitable conditions in their new, impermanent ones—or transportation infrastructures causing degradation and a loss of "spatial health" (Settis 2010, 52; sanità dello spazio) is akin to the positions of those humanistic geographers mentioned by Cresswell who view highways, railways, and airports as "the culprits destroying authentic senses of place" (2006, 31) and, even less surprisingly, to those of many environmentalists. As Jeremy Withers summarizes, quoting Cresswell, "environmentalism has almost always privileged . . . 'a sedentarist metaphysics' . . . that promotes 'place, rootedness, spatial order and belonging (26), and one in which 'mobility is seen as a threat' (42)" (2017, 66). Baldanzi's own kind of "sedentarist metaphysics" emerges from her description of Mugello as a place where there is a stable sense of belonging, with a common dominant culture, language, religion, and ethnicity. In her view, these features are put at risk by the Cemento S.p.A. company and are undermined by various forms of related mobilities and materialities. Remembering the stories generated by this place (including her own) is thus a crucial step to slow down and prevent a complete erasure of such features: "Environmental damages and deaths on the job seem to share the same destiny of impunity and lack of offenders. . . . So, all these stories must not be abandoned" (2011, 273; Danni ambientali e morti sul lavoro sembrano avere lo stesso destino di impunità, di mancanza di colpevoli. . . . Ecco, tutte queste storie non vanno abbandonate).

A vision of mobility as threatening the health of a specific territory also informs Mauro F. Minervino's *Statale 18*. However, given the relevance the author's own car and commuting habits have as agents of knowledge and creative reflection, this work seems to also be characterized by what Cresswell calls a "nomadic metaphysics." This is a conceptual position that does not always "code mobility as negative and threatening" (2006, 42–43) but rather assigns to it a more positive and, I suggest, revelatory value in ecocritical terms.[40] Minervino's book lies midway between a lyrical, disenchanted travel diary and a rigorous geo-ethnographic inquiry—some define it as "ethnofiction"— whose focus and main protagonist is the homonymous SS18, also known as the Strada del cemento (Cement road). This is the long, heavily trafficked artery

(600 kilometers, or 372 miles) that stretches from Naples to Reggio Calabria along the Tyrrhenian coast almost parallel to the A3 Highway (a southern section of the A1). As the narrator colorfully puts it, SS18 is largely responsible for the topographical and "anthropological uglification of the life of the villages and their inhabitants" (Minervino 2010, 17; mostrificazione antropologica della vita dei paesi e degli abitanti raccolti sulla strada) that it sucks into its growing sphere of influence. Echoing Baldanzi's reflections on the TAV and VAV in Mugello, Minervino too is very much aware that the environmental impact of any highway or major road is much larger than the actual strip of land it occupies and that one's construction is frequently just the pretext that anticipates the subsequent degradation of the nearby territory and society. As Minervino writes, "The road is the mother of all the catastrophes in Calabria. The more road there is, the more abuses are born" (34; La strada è la madre di tutte le catastrofi calabresi. Più strada c'è, più abusi nascono).

Implicitly dialoguing with Salvatore Settis's observations in his *Paesaggio, costituzione, cemento*, Minervino, an anthropologist by training, is aware that any environmental crisis is always also a cultural crisis. For him, as for Settis, there is a direct link between spatial and individual health, and whenever private interests prevail over public ones, people and places face dire consequences. Significantly, both authors adopt a medical rhetoric to convey their message: Settis's "spatial health" (2010, 52; sanità dello spazio) becomes Minervino's unstoppable "spreading infection" (2011, 72; infezione [che] si allarga) caused by cement's and asphalt's viral impact on the environment. The crucial idea of *Statale 18* is that, as Minervino puts it in one of its clearest formulations, "caring for the interior requires an attention to the exterior: most of our soul is outside our bodies. . . . The interior is everywhere: wherever we look and listen. [It is] on this road too" (2010, 98; la cura per l'interno richiede attenzione per l'esterno; la maggior parte dell'anima sta fuori del corpo. . . . L'interiore è dovunque: dovunque guardiamo e ascoltiamo. Anche su questa strada). This remark further corroborates the unavoidable interconnections between private and public bodies, biological and sociopolitical processes, spaces and places, matter and meaning—or, quoting Karen Barad, the *"mutual constitution of entangled agencies"* (2007, 33).

In Minervino's account, the SS18 is compared to a "sharp skewer that pierces and runs through the human and social flesh of the main centers overlooking the Tyrrhenian coast one after the other" (2010, 14; uno spiedo appuntito che trapassa e infilza tutta la polpa umana e sociale dei centri principali che stanno affacciati uno dopo l'altro sulla costa Tirrenica). It progressively transformed over the last few decades into an "urban highway" (superstrada urbana) and an endless "urbanized corridor" (18; corridoio urbanizzato), actively intermingling with people's lives; it is a "document and the reflected gaze of what we are" (83; è insieme documento e sguardo riflesso di ciò che siamo). In short, this road "simultaneously follows life and determines it" (56; insegue la vita e insieme la determina) and, as such, emerges as a Latourian hybrid constituted by a network of interrelated agencies. It is a constantly shifting, chaotic assemblage of human and nonhuman elements stickily flowing together and variously affecting bodies, relationships, and subjectivities.

Antipodal to its name, we may also add that this road is the ironic emblem of an almost total absence of the state in this part of Italy, especially if one understands the latter, in both its national and regional ramifications, as an institution that should provide some guidance and rules regarding the implementation and control of building regulations, lawfulness, or respect for the environment. In this sense, the SS18 itself, together with the degradation of the surrounding territory it facilitates, might be seen as a material manifestation of the peculiar pathologies characterizing this part of Southern Italy and Calabria in particular. This is a region where the state has historically been perceived as hostile and oppressive and frequently unable (or unwilling) to respond to people's needs. Thus citizens have turned to local criminal organizations. That said, the author encourages us to look at the SS18 as stretching beyond its geographical limits. Indeed, it is emblematic of a widespread economic, social, cultural, and (encompassing them all) environmental crisis in Italy and elsewhere. This is a crisis that appears wherever political and criminal interests coalesce to produce "the illusion of development, but actually only [to] destroy resources and mortgage one's future" (Minervino 2010, 31–32; l'illusione dello sviluppo, ma in realtà distrugge solo risorse e ipoteca il futuro).[41]

Like most of the discursive and material texts examined in this book, the SS18 thus becomes a sort of laboratory and portable, universal interpretative key that crosses specific (regional, national) borders. Heeding Heise's observations in her *Sense of Place and Sense of Planet* (2008), the SS18 in turn helps us individuate and decipher similar global landscapes that become overwhelmed and degraded whenever the ideas of modernity, progress, and socioeconomic development are associated with material construction and the circulation of capital, on the one hand, and a contempt for memory, nature, and culture, on the other. As Minervino eloquently puts it, SS18 is one among those slippery configurations of modern living that identifies at once with the "Western sprawltown," and with formations typical of the third and fourth world, such as the megalopolises constituted by the accumulations of *slums* or *barrios*.

una di quelle configurazioni incerte dell'abitare nella modernità che si identificano nella "sprawltown occidentale" e nelle megalopoli costituite prevalentemente dall'accumulo di *slums* o *barrios*, formazioni tipiche del terzo e del quarto mondo. (2010, 34)

Later on, in a radio interview with Angelo Ferracuti, Minervino adds that "mobility has become the predominant feature for Calabrians . . . paradoxically, they are more modern than the Chinese" (2011). These words are quite distant from what Corrado Alvaro (1895–1956), one Calabria's most popular writers of the past century, wrote in his *Itinerario italiano* (Italian itinerary) when he reflected on his native region's traditional transportation structure. Alvaro recalls an image from his childhood of an underdeveloped "Calabria where one used to travel on foot or on the back of a mule" (Calabria che si percorreva a piedi o sul mulo) and how, a few years later, one could "travel far and wide on some of the most beautiful roads in Italy" (1995, 226; percorrere in lungo e in largo con le strade tra le più belle d'Italia).

Although this could still be sporadically realistic and truthful, especially away from densely populated coastal areas, there are few doubts that much of Minervino's Calabria around SS18—hypermobile, overbuilt, and automobile-obsessed—is closer, both spatially and temporally, to the spreading Italian northwest, to suburban Los Angeles, or to the smoggy peripheries of Lagos and Shanghai than to a stereotypically slow-paced, citrus-scented Southern Mediterranean landscape where "man lives still subdued in the middle of nature, as if he were close to a beast of whom he does not know the strength" (Alvaro 1995, 217; l'uomo vive in mezzo alla natura ancora sottomesso, come presso una bestia di cui non conosce la forza ma che sa potente).[42]

At this point, what is even more tempting is to consider Minervino's remark on the mobility of Calabrians in the context of his observations about the interrelations between the outside and the inside and on the ways in which human beings—their bodies and minds—are deeply affected by their neglected environment. In other words, the situation is one in which "architecture, the shape of built space, is together a document and a reflected gaze of what we are" (2010, 82; l'architettura, la forma dello spazio costruito, è insieme documento e sguardo riflesso di ciò che siamo). Similar to the TAV and VAV dynamics in

Mugello, this augmented mobility ostensibly afforded by the SS18 and related infrastructures does not correspond to any freedom of movement, existential well-being, or decreased sense of disorientation on the part of the typical local resident. On the contrary, the resident (usually an adult male) is depicted as either feeling "trapped in [his] car" (63; intrappolat[o] in macchina), "driving a SUV, and living and building his home on the *new road* as if he were still bogged down in the same misery, the same enclosed remoteness of his disowned ancestors" (64; gira magari col suv, ma che vive e costruisce la sua casa sulla *strada nuova* come fosse ancora impantato nella stessa miseria, la stessa lontananza recintata degli avi ripudiati), or ready to emigrate "to study and live elsewhere" (64; per studiare e vivere altrove). The storied network constituted by territory, misguided infrastructural planning, flawed self-serving politics, and material substances constituted by the SS18 thus generates a chain of socioenvironmental disasters.[43]

For example, it may be no longer striking to learn that "Calabria today is the Italian region with the highest rate of illegal building" (36; la Calabria è oggi la regione italiana a più alto tasso di abusivismo edilizio) or that "less than two million residents in this region could contend for ... more than eight million rooms" (34; meno di due milioni di calabresi sparsi per la regione, potrebbero contendersi ... più di otto milioni di stanze) while empty interior villages count scores of dilapidated structures that nobody cares for. However, it is probably more surprising to realize that despite all the construction frenzy and lodging availability, when some segments of the population (students, low-income individuals, and migrant agricultural workers) actually need a house, they cannot find one they can afford and, as a consequence, are either stuck in place or compelled to emigrate. What emerges from the sprawling SS18 transportation ecosystem is thus both a physical and metaphorical "immobility"—that is, a sense of resignation and isolation among some of the people who live around it—and a twisted sort of mobility that is infused with "prejudice and inequality" (Withers 2017, 68) and is linked to exploitation, as it arguably compels some to leave Calabria and perhaps find work in Mugello's TAV or VAV tunnels.

As I anticipated at the beginning of this chapter, however, mobility in *Statale 18* is not exclusively viewed in negative terms. In fact, Minervino implies that a "sedentary metaphysics" (Withers 2017, 26), as environmentally attractive as it may be, not only is no longer a viable and concrete option today but may also constitute an impractical dead end in environmental and ecological terms. In what appears to be a paradox, the same network of mobile transportation infrastructure and technology (humans, roads, cars) that gives form and substance to the devastated SS18 ecosystem is also what enables one to analyze its shortcomings and envision alternative, more sustainable scenarios. In other words, the human-automobile combination of the narrator and his car, as he drives along the SS18 to commute to work, may be interpreted as one of the assemblages mentioned by Withers that works to "foster a deeper knowledge of the intricate behavior of nonhuman life" and to glean "a more nuanced understanding of the features and processes that shape and define the desert [here literal and metaphorical] around him" (2017, 71–72), guiding us in turn toward forms of resistance that are not just political but also aesthetic, sentimental, and moral.

In this sense, three sections of the book, "Dromofilia" (Dromophilia), "Strade azzurro cupo" (Dark-blue streets), and "La villa delle ragnatele, i tonni" (The spiderweb villa, the tuna fish), are especially revealing. On the one hand, they further clarify the protagonist's perception of such prosthetic human-machine assemblage—through words, one could jokingly add, that would not have displeased Marinetti: "The cabin is transformed into an interior space, larger than [my] body . . . the car is almost an extension of my home, a shell that I bring along" (Minervino 2010, 199; L'abitacolo si trasforma in uno spazio interiore, più vasto del corpo . . . l'automobile è quasi un prolungamento del mio domicilio, un guscio che mi porto dietro). This human-machine hybrid enhances Mauro's perceptive skills: "[I] am in a condition of abandoned vigilance that fosters detachment, revisions, unintentional anamnesis . . . on SS18 I reflect on many things" (107, 199; Uno stato di abbandono della vigilanza che favorisce il distacco, le revisioni, l'anamnsesi involontaria . . . sulla 18 rifletto su molte cose). On the other hand, these pages also illustrate how well such an assemblage may function as a cognitive tool that reveals the complex behavior of nonhuman life and the extent to which the latter is deeply interrelated with our own. Thus, for example, as the driver's attention is caught by some of those forgotten "landscapes in between" (Seger 2018) along SS18 that lie beyond old guardrails or new concrete barriers, he observes first how such interstitial spaces provide shelter to the many organisms— (named "saprophytes" to indicate insects, magpies, seagulls, worms, snakes, and rats)—whose own lives, similar to that of many "profiteering" humans, depend on and take advantage of this road. Next he observes "the animals killed and crushed every day by the pointless butchery caused by cars" (Minervino 2010, 206; le bestie abbattute e stritolate ogni giorno dall'inutile macelleria delle macchine). Then in the following paragraph, a powerful juxtaposition reminds us that in this sort of ecosystem, human beings are actually roadkill or, better, "snarge" (Kroll 2018) as well and that in these circumstances, all hierarchies and strict ontological distinctions become questionable:

> The street is a level, it knocks down and lowers even the memory of the dead at the fatal height of the dirty asphalt. All are at this ground level, the one of the smashed dogs, of the ants and the mice.
>
> La strada è una livella, abatte e abbassa anche i ricordi dei morti all'altezza fatale dell'asfalto sporco. Tutti al livello del suolo, dei cani sfracellati, delle formiche e dei topi. (Minervino 2010, 206)

A few pages later, as he imaginatively overlaps via a lyrical similitude the SS18 with a sea-lane traveled by tuna fish, the protagonist's ideas of the SS18 as a sort of "level" that reduces ontological distances become even more explicit:[44]

> When I drive on the state road, I push ahead like tuna fish. I follow the sea-lane traced by tuna. Tuna, like human beings, have long bones and warm blood, they live by swimming thousands of miles every year and, by now, they are hardly surviving, [finding themselves] in different and always more difficult environmental conditions. Like me they push ahead always parallel to the coast, never too far from the shore. . . . Like human beings, tuna have an instinctive sense of orientation, memories of place and of life circumstances. Tuna always swim . . . and they always find their path. . . . I often think of the tuna fish that swim just under the water's surface, and I always turn to the side of the sea that runs flat . . . to starboard of my car window.
>
> Io quando guido sulla Statale vado avanti come i tonni. Seguo la rotta dei tonni. I tonni come gli uomini hanno le ossa lunghe e il sangue caldo, vivono nuotando migliaia e migliaia di chilometri ogni anno e sopravvivono ormai a stento, in condizioni ambientali differenti e continuamente più difficili. Come me vanno avanti sempre parallelamente alla costa, mai troppo lontani da una riva. . . . Come per gli umani, istintivo nei tonni sono il senso dell'orientamento, la memoria dei luoghi e delle circostanze di vita. I tonni nuotano sempre . . . e nuotando sempre trovano la loro strada. . . . Penso spesso ai tonni che nuotano sotto il pelo dell'acqua e resto sempre girato dal lato del mare che fila piatto . . . a dritta del mio finestrino. (Minervino 2010, 215–16)

The fact that human beings and tuna fish have an intertwined destiny is almost too obvious to point out: they are both at risk of extinction if, literally and figuratively, they lose their orientation and their habitats are progressively degraded. However, there may be a less pessimistic significance in the imaginary human-animal encounter described in this passage if, heeding Roberto Marchesini, one recalls that "the animal epiphany . . . becom[es] the herald of new existential dimensions . . . able to disclose contents and contradictions" (2014, xxviii). In other words, as we approach the end of a section and of a road where not only cement and asphalt but also the conceptual tools of a "sedentarist" and "nomadic" metaphysics have played a relevant role, the evocation of a fish that is the epitome of speed and mobility and yet, like us, is also aware of its location and has "memories of place" suggests its interpretation as both a vehicle and an arche—"the refuge, the dwelling of humans within the space of nature" (Minervino 2010, 81; il riparo, la dimora dell'uomo dentro lo spazio della natura)—that is, as a hybrid fantastical being that is ultimately able to "break open the human system and introduce new, restructuring seeds" (Marchesini 2014, xxviii).[45]

It would be an impossible task to identify all the "restructuring seeds" introduced in this specific context. However, if one considers that these tuna fish are at once real and imaginary literary creatures, we might argue that among those seeds, there is also the story that contains them. Or, to underline what should be a familiar point by now, similar to the tuna fish, Minervino's awareness-raising and ethically informed narrative is in itself a seed planted in the reader's mind, an arche reminding us of lost beauty, and an inspiring vehicle for change and transformation. As Minervino has stated, "Literature and poetry always enlighten utopian but also possible planes of reality" (2011; La letteratura e la poesia illuminano sempre un piano di realtà utopiche ma possibili). As *Statale 18* documents and counters the contempt for culture and nature that the SS18 emblematizes—eventually dissolving concrete, bricks, and asphalt into an uncluttered liquid highway populated by moonlit fish-humans—it transports us to "new ecologies . . . of words, of ideas . . . of new possible material realities" (Iovino 2016a, 4) and invites us to reimagine its material textuality in more joyful and environmentally sensible ways.

At the conclusion of this chapter, it does not hurt to emphasize once again cement's and asphalt's active roles in tangibly shaping territories, communities, and imaginative practices and perceptions. These materials are similarly central to Expo 2015, a "concrete" megaevent designed for an ephemeral public use; to the fictional, ambitious home remodeling project in Sarchi's novel; and finally, to the "permanently expanding" transport infrastructure in Minervino's literary "road documentary."

While aggregates such as concrete and asphalt implicitly denounce, in all these different yet related instances, systemic imbalances and the dystopian hazards caused by cultural and ethical lacunae, they also speak of their own "aggregation" with human beings. Concrete conveys humanity's undaunted (yet also pathetic) aspiration to leave its mark and matter more than anything (and anyone) else while itself being "a powerful force . . . [that has] allowed the constant acceleration of modern lifestyles" (Harkness, Simonetti, and Winter 2018, 37–38).

Epilogue

As this ecocritical exploration around the Italian Peninsula through texts, places, and substances comes to an end, there should hopefully be few remaining doubts about the onto-epistemological vitality, semiotic wealth, and expressive potential of the materials explored in these pages. Rather than simply considering sulfur, marble, oil, steel, asbestos, cement, and asphalt as literary or visual tropes, inert substances, or mere resources for consumption and commodities, *Elemental Narratives* has made every effort to view them as intimately enmeshed and entangled with human beings, endowed with narrative agency, reciprocally involved in transforming our surroundings, and shaping a world that, as Jane Bennett writes, is "an interfolding network of humanity and nonhumanity" (2010, 31).

Taking advantage of recent ecomaterialist theories within the environmental humanities that expand the notion of agency beyond the human and redistribute it to include a plurality of nonhuman actors, we viewed matter in general and the above-mentioned substances in particular in a relationship of mutual interdependence and interaction with human beings and as dynamic cocreators and coconveyors of "configurations of signs and meanings that can be interpreted as stories" (Oppermann 2017, 293).[1] Although distant in place and time and informed by different ideologies, the specific narratives selected in the five chapters of this volume share at least one of the following characteristics: an aesthetic and ethical interest in place, a tendency to question strictly anthropocentric perspectives, and a more or less explicit desire to spark reflection on and awareness of the increasing challenges (ethical, ecological, technological, cultural, existential) that contemporary Italy shares with the rest of a planet in crisis—on the verge of the "sixth extinction" (Kolbert 2014).

The literary, visual, and material texts I have discussed reveal diverse forms of distributed agency and natural-cultural entanglements between bodies and landscapes, the human and the more-than-human, and the organic and the inorganic while offering situated perspectives on salient socioenvironmental concerns and challenges across Italy and the Mediterranean. At the same time, they invite us to individuate and reflect on potentially comparable situations elsewhere in the world, wherever and whenever there is the risk, as Timothy Morton puts it, to "squander our inheritance, fail to acknowledge our debts, forget the bodies and the materials that have made us what we are" (2010, 41). Last but not least, as many of these Italian-Mediterranean narratives necessarily insert themselves within the broader discourse of the Anthropocene, they question essentialist and anthropocentric world visions and critically reappraise humans and their role on the planet in a broader posthumanist context—one where, for better or worse, we are inextricably bound with others, "including the nonhuman or 'earth' others" (Braidotti 2013, 48).[2]

As we have seen, the acknowledgment that human bodies are commingled with their environment has far-reaching cultural, environmental, and sociopolitical implications. Countering those tendencies that advocate for the separation of mind and matter or support fundamentalist and nonpluralistic world visions, this book is premised on the belief that our lives and destinies are intertwined with those of other entities, both organic and inorganic. In a recent article in the *New Yorker* entitled "The End of Sand," journalist David Owen provides clear and convincing evidence of this sort of interconnection while showing that the idea is increasingly gaining ground beyond the academic sphere. Referring to different sand-extraction sites worldwide (Lake Ontario, Connecticut, the Gulf Coast, Dubai, and the New Jersey coast after Hurricane Sandy), he begins by stating that sand "is one of our most widely used natural resources, but it's scarcer than you think" (2017, 28). Owen concludes by implying that if and when the last grain is gone, some of us may well be gone too—swept away by an "ocean [that] has risen too high to be held back by sand" (33). The validity of this argument is, of course, transferable to additional substances, places, and circumstances.

Despite the Italian Peninsula's 4,720 miles of coastline, my own exploration only includes sand in some passing references. Sharing Owen's environmental concern, I rather discuss sulfur mining's effects on people and places in Sicily in some of Luigi Pirandello's classic short stories, marble's lethal economic power in the Apuan Alps, petroleum's responsibility in developing corporeal and communicative diseases in Sardinia (and the role that a fact-based documentary and a work of fiction have in defining such a material responsibility), asbestos's effect on bodies and landscapes in Alberto Prunetti's *Amianto* and its toxic legacy in Casale Monferrato, and concrete's and asphalt's socioenvironmental impact in Mugello and Calabria. My study investigates entwined assemblages of matters and humans and combines "thinking the limits of the human with thinking elemental activity and environmental justice" (Cohen 2015, 5).

The idea that there are connections among different forms of exploitation is also part of the foundational logic of this book. If matter runs the risk of being treated as just "a reservoir of tractable commodities," objectified human beings face the parallel danger of being considered "expandable resources: miners who can be discarded once they develop black lung, or minority communities that can be tallied, televised, and toured after a hurricane obliterates homes" (Cohen 2015, 5). The overall health of our collective future also depends on the extent to which we are willing to recognize such a situation of shared vulnerability while developing what, borrowing the expression from Herbert Marcuse, one could call a "new sensibility" (Farr 2019) that is conducive to transforming an ever-less sustainable status quo into a more equitable one.

In his introduction to *Elemental Ecocriticism*, Jeffrey Cohen writes that "attending to matter and writing against the reduction of world to commodity (resource, energy) is a powerful aid to activism" (2015, 4). I will be gratified if *Elemental Narratives* contributes, even minimally, to such a purpose or if it helps us imagine the potential advantages of a posture of universal respect and care for both the human and the more-than-human—if, in other words, this work might help articulate "a grammar that is different from the one of expansion and power" (Cassano 2012, lv). As indicated throughout, the process of accelerated growth and industrialization that affected Italy in the past decades can be tracked in several other locations globally. Enriching our understanding of the relationship between humans and their environment in a country such as Italy, beautiful yet full of conflicts and contradictions, where history permeates the landscape and the blend of environmental, sociopolitical, and cultural dimensions has an almost tangible quality is not just an effort for its own sake. Rather, it provides a model for other scholars who are interested in the interrelations of the local with the global, in extending ecocriticism's range and methodologies beyond more frequently discussed geographic (and limited academic) borders, and in bringing out similarities and differences between environments that may be affected by comparable problems—be they toxic contaminations, unnecessary megainfrastructural projects, or aggressive extraction policies. As the transdisciplinary field of the environmental humanities continues to grow, offering more ways to respond to catastrophic changes and cope with the many ecological crises our planet is currently facing, this contribution—bioregionally focused yet also hopefully "free of the bridles . . . of nationalism" (Verdicchio 2016, xv)—continues the dialogue between ecocriticism and Italian studies while also engaging with larger, transnational conversations.

At a past Association for the Study of Literature and Environment (ASLE) conference, while still gathering ideas for this project, I was heartened to hear that the (post)humanities "reconceived" must deal not only with spirit but also with matter and that certain narratives may foster an art of noticing that can have positive environmental effects by slowing down the current practices of extinction. While I am confident that the narratives discussed in this book possess such an enlightening and awareness-raising quality, I wish to conclude with an excerpt from a text that manages to synthesize in a few words much of what I have been trying to say all along in this volume:

> An allegedly universal discourse must be at the same time the discourse of metals and salts and of rocks, of beryllium, of feldspar, of sulfur, of rare gasses, of the nonliving matter that constitutes almost all of the universe.

> Un discorso che si presuma universale dev'essere insieme il discorso dei metalli e dei sali e delle rocce, e del berillio, del feldspato, dello zolfo, di gas rari, della materia non vivente che costituisce la quasi totalità dell'universo. (Calvino 1994, 1158)

In the previous chapters, I have attempted to let the Italian voices of some matter and places speak alongside the creativity of writers, filmmakers, and artists. My goal has been to argue, along with Calvino, that only this sort of horizontal, encompassing, and quintessentially posthumanist listening exercise may increase our chances of reconfiguring perspectives and cultural imaginaries. In these voices and in these stories too, we might look for sensible solutions to an ever more urgent and universal socioenvironmental crisis.

Notes

Introduction

1. When available, I use the published English translations of the original Italian texts (included in the references). All remaining translations are mine.

2. Levi writes, "This cell belongs to . . . my brain, to my writer's brain, and this cell at issue, and with it, the [carbon] atom at issue, is assigned to my writing, in an enormously minuscule game that nobody has described yet" (1984, 232; Questa cellula appartiene [al] . . . mio cervello, di me che scrivo, e la cellula in questione, ed in essa l'atomo in questione, è addetta al mio scrivere, in un gigantesco minuscolo gioco che nessuno ha ancora descritto).

3. For an articulated and in-depth analysis of these aspects in Levi, see Benvegnù (2018).

4. Here I am referring to Iovino and Oppermann's essential point that matter tells its stories "through the imagination of human counterparts" (2014, 82). Climate change, the recognition of the complexity of the makeup of human bodies, and worldwide immigration crises could, respectively, exemplify each of the cases mentioned.

5. Iovino notes that "more than other countries . . . Italy is almost inevitably synonymous with its landscape" (2016a, 3).

6. James writes that her approach "is not limited to postcolonial texts" (2015, 24). Within Italian studies, Sandra Ponzanesi observes that postcolonialism should "be understood . . . not as a chronological transition from a colonial to a postcolonial status, but as a theoretical tool that aims to critically assess the operations of empires and their lasting legacies and effects in present day society" (2012, 59).

7. The "Seveso disaster," which is considered the worst environmental accident in Italy, happened on July 10, 1976, when a highly toxic dioxin cloud came out of the ICMESA (Industrie Chimiche Meda Società Azionaria) chemical plant situated in the Lombard village of Seveso and contaminated a large area nearby. Twelve years later, almost on the same date, July 17, 1988, a fire caused the escape of another cloud full of toxic chemicals used for agricultural insecticides from the Farmoplant plant in Massa Carrara. It polluted more than seven hundred square miles in the regions of Tuscany and Liguria.

8. See, for example, Alaimo and Hekman (2008); Coole and Frost (2010); and Dolphijn and van der Tuin (2012).

9. Iovino (2016b) discusses a posthuman ethics based on the coextensive materiality of subjects, both human and nonhuman, and lists some of the most influential studies on the subject.

10. As Posman, Reverseau, and Ayers write, "Modernist and avant-garde art and writing often focus on the things that surround people, the houses and cities we live in, the bodies we call ours, the sensations we experience and the words our consciousnesses spill" (2013, 3).

11. Cannamela (2019) takes an excellent step in this direction.

12. Meschiari (2008, 48–49) reprises Bateson's classic idea of an extended mind that pervades nature as well as culture and addresses some Italian correspondences between "landscape" and "mindscape."

1. In her discussion on how Anglo-American modernism and ecocriticism can be productively engaged, Raine points out that as early as 1998, Cantrell argued that modernist texts are relevant for ecocriticism because modernist writers are self-conscious "witnesses to the profound changes in the human relations with the planet that ha[d] become visible in [their] century," sharing a sense of having experienced a "revolutionary change" in "'the given' we call nature" (Raine 2014, 101).

2. Falci continues by stating, "In the postwar period, such a mode has intriguing historical implications. . . . For poets from decolonizing places, this mode becomes a means to articulate aesthetic independence from imposed imperial norms and to revalue native experience while drawing on and experimenting with models of European modernism and the English literary tradition. For poets in Britain and in England in particular, such an elemental poetics has quite different ramifications" (2015, 65).

3. In the third and last section of the book, which corresponds to the protagonist's adulthood and poetic maturation, the city becomes a place "from which it is impossible to disentangle oneself" (Coda 2002, 167).

4. Bond notes that "it is only when he acknowledges both sides of his personality (thus fusing his natural, instinctual soul with his higher spiritual aspect, as represented by the poet Pennadoro) that he feels ready to descend among the city-dwellers" (2016, 7). Pennadoro is the Italianized form of the surname Slataper (in Czech, *zlato* means "oro" [gold], and *pero* means "penna" [pen]).

5. As the protagonist states toward the conclusion, "Nature, I thank you. You made me free, and I thank you" (Slataper [1912] 1988, 144; Natura, io ti ringrazio. Tu m'hai fatto libero, e ti ringrazio).

6. On the presence of sociopolitical, anticapitalist concerns in Pirandello's work, see Cesaretti (2013).

7. Sulfur is not included in Pinkus (2016).

8. As Consolo observes, sulfur's "dramatic epopee . . . reached its apex between the end of the nineteenth and the beginning of the twentieth century, and then decreased until its disappearance in the '50s" (1999, 198; drammatica epopea . . . raggiunge il suo acme tra la fine dell'Ottocento e gli inizi del Novecento, decresce fino a sparire verso gli anni '50). After deposits were discovered in 1867 in Louisiana and Texas and a new extraction procedure was invented (the "Frasch method"), American sulfur became more competitive and eventually replaced Sicilian sulfur in world markets.

9. In "Ciàula scopre la luna," one finds a similar image. In this instance, the narrator further clarifies the relationship between the dismal situation of the landscape and the sulfur miners' bodies:

> In those hard faces almost faded by the raw darkness of the underground caves, in those bodies exhausted by the daily labor, in those torn clothes, they had the livid squalor of those lands without a grass blade, pierced by the sulfur mines, similar to many, enormous anthills.
>
> Nelle dure facce quasi spente dal bujo crudo delle cave sotterranee, nel corpo sfiancato dalla fatica quotidiana, nelle vesti strappate, avevano il livido squallore di quelle terre senza un filo d'erba, sforacchiate dalle zolfare, come da tanti enormi formicaj. (Pirandello 1994a, 1279)

10. Toward the end of "Il fumo," sulfur becomes ubiquitous: "One could not see anything but sulfur everywhere in that village! Sulfur was also in the air one breathed, and it took off one's breath, and burned the eyes" (Pirandello 1994d, 77; Non si vedeva che zolfo, da per tutto, in quel paese! Lo zolfo era anche nell'aria che si respirava, e tagliava il respiro, e bruciava gli occhi).

11. For an in-depth discussion of extractivism as "the dominant paradigm of contemporary capitalism and neoliberalism at large," see Mezzadra and Neilson (2017, 1–21).

12. Pirandello had firsthand knowledge of the problems and concerns related to sulfur. His father owned a mine that eventually

caused him substantial financial losses. As Don Mattia Scala remarks, "The devil damns your soul but, if he wishes, also makes you rich! Sulfur instead makes you poorer than Saint Job, and still damns your soul!" (Pirandello 1994d, 48; Vi danna l'anima, il diavolo, ma vi fa ricchi, se vuole! Mentre lo zolfo vi fa più poveri di Santo Giobbe: e l'anima ve la danna lo stesso!).

13. Other events include the sudden departure of his only son, a pet monkey killing its owner by digging a hole in the owner's throat with its nails, and Don Mattia's failed negotiations with other landowners, lawyers, and mining engineers as well as his progressive sense of isolation and alienation.

14. To counterbalance such a dark reference to trees, we might recall a "greener" passage in *One, No One, and One Hundred Thousand* in which the protagonist, Moscarda, expresses a vitalist urge to become one with the outside world: "This tree, tremulous breath of new leaves. I am this tree. Tree, cloud; tomorrow, book or wind: the book I read, the wind I drink. Everything outside, wandering" (Pirandello 1992, 160; Quest'albero respiro trèmulo di foglie nuove. Sono quest'albero. Albero, nuvola; domani libro o vento: il libro che leggo, il vento che bevo. Tutto fuori, vagabondo).

15. I agree with Viola when he observes that Marinetti's obsession with matter aims at liberating

> art from another obsession, the anthropomorphic one, and in particular, allows the artist to free him/herself from the "I" of the author. By liberating the work of art from the authorial burden, the former may dedicate itself to its true objective, namely to weave analogical webs; only such webs will be able to capture matter which, within the sea of phenomena, is alchemically pervaded with life.
>
> l'arte da un'altra ossessione, l'ossessione antropomorfica, e soprattutto consente all'artista di liberarsi dall'io dell'Autore. Liberando l'opera dall'ipoteca autoriale, ecco che essa si può dedicare al suo vero scopo, che è quello di tessere reti di analogie; sono solo queste reti che consentiranno di catturare la materia, che è quasi alchemicamente tutta pervasa di vita, nel mare dei fenomeni. (2010, 4)

16. This point is reaffirmed, though from a different perspective, in Daly (2013).

17. In his discussion of futurism's manifesto, Campbell recurs to Foucault's concepts of biopolitics, biopower, and milieu to investigate the idea of the "aliveness" associated with the animal (as juxtaposed to the "presence of death" associated with the human) and provides further proof of technology's essential role (not to mention the importance and agency of the nonhuman animal) in "moving human beings towards a more animalized and hence more vital existence" (2009, 166). He concludes, "Marinetti's biopolitics vitalizes life by incorporating death within it" (171).

18. Blum observes that especially when such an environment or object is perceived as hostile, "the assimilative strategy of metaphorically incorporating [it] is the precondition for artistic creation" (1996, 134).

19. As he writes about futurist "aeropoesia" (aeropoetry) and "aeropittura" (aeropainting), Mangano points out that "air is not only an element in the landscape or a new space for a new kind of landscape, but it became matter itself.... The 'I' must not affirm himself and tyrannically dominate the world.... [In Buzzi] one finds the lesson of Balla and of early futurism: act on the landscape, create this art which transforms the landscape, act directly on landscape and space.... [This is] an attitude that, later on, will bring direct interventions on landscape, such as Land Art" (1996, 62–63). Mangano also observes that Marinetti, in his "L'aeropoema del Golfo della Spezia," does not completely succeed in creating a complete fusion among the machine, the poetic "I," and territory because of the superiority and loquacity of the narrating subject vis-à-vis a newly objectified world (61).

20. *Material-discursive* refers to material ecocriticism's consideration of matter as a text in itself: "Reality is read as a material text, *as a site of narrativity*, a storied matter.

If discourse and meaning are co-extensive in the constitution of matter, reality can be discovered as an array of stories" (Iovino 2012a, 57).

21. On the relationship between futurism and posthumanism, see Terrosi (2009).

22. The following excerpt from the "Il manifesto tecnico della pittura futurista" (Technical manifesto of futurist painting) should be added here:

> Our new conscience does not make us consider man as the center of universal life any longer. For us, the pain of a man is as interesting as the one of an electric lamp, which suffers, and agonizes, and cries with the most excruciating expressions of pain.
>
> La nostra nuova coscienza non ci fa più considerare l'uomo come il centro della vita universale. Il dolore di un uomo è intressante, per noi, quanto quello di una lampada elettrica, che soffre, e spasima, e grida con le più strazianti espressioni di dolore (quoted in Terrosi 2009, 271).

23. I realize that the kind of embodiment I am describing here may have some affinity with the definition of *personification* mentioned by Ceccagnoli—namely, "when a nonhuman entity is transformed into a human sentient entity, able to think and speak, and has a voice and a face" (2009, 320). I am suggesting that in some of his texts, Marinetti's intention is to achieve something that moves even beyond personification. As Bennett notes, "This vibrant matter is not the raw material for the creative activity of humans or God. It is my body, but also the bodies of Baltimore litter . . . Prometheus's chains . . . as well as the not-quite-bodies of electricity . . . ingested food . . . and stem cells" (2010, xiii).

24. In addition to the scholars mentioned so far, Curi (2009) dealing with linguistic issues and, more recently, Delville addressing questions of alimentary aesthetic have discussed futurism's materiality of language and the "avant-garde . . . general engagement with the physicality of concrete and textual matter" (2008, 127). Focusing on some of futurism's erotic short stories, I have previously emphasized a futurist drive toward an inorganic and yet sentient sexuality building on Perniola's observations in his *The Sex Appeal of the Inorganic* (2004); see Cesaretti (2013).

25. In *Venezianella e Studentaccio* (2013), Marinetti will reprise and conclude his lifelong poetic elaboration on Venice. In his introduction, Paolo Valesio notes that the novel "expresses . . . the desire to stabilize Venice, protecting it from the danger of sinking" (2013, lxviin25), thus also indicating that some kind of environmental sensibility is not completely alien from Marinetti. In his *Manifesto dell'architettura futurista* (Manifesto of futurist architecture), then, Antonio Sant'Elia notes that "for architecture, one must intend the effort to harmonize the environment and man with freedom and great boldness—that is, to make the world of things a direct projection of the world of the spirit" (1914; per architettura si deve intendere lo sforzo di armonizzare con libertà e grande audacia l'ambiente con l'uomo, cioè rendere il mondo delle cose una proiezione diretta del mondo dello spirito).

26. Here are my very literal translations of some of the characters' simultaneously unnatural and nature-echoing names: Vasto/Vast, Altaluce/Highlight, Alata/Winged, and Ballamar/Dance-sea.

27. The epigraph in Palazzeschi's futurist poem *L'incendiario* (The incendiary) reads, "To F. T. Marinetti the soul of our flame" (1910, 69; A F. T. Marinetti anima della nostra fiamma) and captures Marinetti's energizing role within futurism and his disruption of past poetic traditions.

28. Another instance of technological failure is in the fifth synthesis when a "faraway voice" (Marinetti 1960d, 186; voce distante) announces that the cable car used to bring ice and snow down from the top of the mountain has fallen due to the volcano's imminent eruption.

29. As Valesio suggests, poetry might be a kind of ecological site of its own: "The evocation of poetry as a privileged and ideal space, a haven from the ugliness of the world,

traverses all of Marinetti's work, from its pre-futurist (i.e., symbolist) phase . . . to its futurist one, despite the various desecrating moves" (2013, lxvn17).

Chapter 2

1. This essay is a review of the Jordanian-Saudi writer Abdelrahman Munif's *Cities of Salt*, which Ghosh describes as a "monumental five-part cycle of novels dealing with the history of oil" ([1992] 2005, 142).
2. In 2016, oil fields were discovered near the Tremiti and Pantelleria islands (off the coasts of Puglia and Sicily, respectively) and sparked a heated national debate about the logic and opportunity of drilling in areas of such natural beauty.
3. On the history of oil exploration in Italy, see also ENI's website at http://www.eni.com.
4. In "The Creative Anxiety of the Hydroelectric Plant Nera Velino," futurist artist Franca Maria Corneli celebrates precisely this kind of achievement by referring to the spectacle offered by the plant built at the confluence of the Nera and Velino Rivers, between the regions of Marche and Umbria: "The furor of your precipitating waters in the sonorous metallic tubes down always down in the galleries reverberates its spasmodic and tortured joy from valley to valley all the way into the distant ravines of the Apennines with their thousands and thousands of echoes as an audience" (quoted in Rainey, Poggi, and Wittman 2009, 472).
5. Significantly, in his October 30, 1926, speech in Reggio Emilia, Benito Mussolini echoes Marinetti's vision when he states, "It is imperative that we create; we, people from this epoch and this generation, because we have the duty to remake the face of the Fatherland both spiritually and materially. In ten years, comrades, Italy will be unrecognizable! This is because we will have transformed it, we will have made a new one, from the mountains which we will have covered with a green coat, to the fields which will be completely reclaimed" (2012).
6. Armiero notes that "in theory, the reclamation of the land would lead to the reclamation of people, in body and soul. In reality those poor internal immigrants paid a high price for the Fascist reclamation scheme" (2014a, 241).
7. Ethiopia was an Italian colony between 1936 and 1941. Additional references made to a "child" who "draws over there / that poor hut / that smokes straight / on the white page of the horizon" and to a "sheep that fled / from the child's crèche / grazes and does not hear [the song of petroleum and the noise of the drill]" (Goretti [1941] 2009, 477) reinforce the stereotypical image of a poor, rural African landscape where humans and animals live in close proximity. Orban's observation that "after the Ethiopia campaign of 1935, the Allies blockaded oil shipments to Italy as a punishment, as Italy was almost entirely dependent on foreign oil" (quoted in Pickering-Iazzi 1995, 68) supports this geographical hypothesis.
8. The following verses illustrate petroleum's perspective from below: "Light . . . / You don't know it Man / But every morning / she came to meet you / every evening / she watched over your sleep / She has kissed your eyes / luminous kisses / and you have seen beauty / I haven't / I a prisoner / The mountain above me" (Goretti [1941] 2009, 478).
9. As hinted earlier, Marinetti's relationship with the city is not simple or straightforward. In what may be interpreted as a preservationist, nostalgic dream of "a city that does not grow old" (città che non invecchia), he aimed at constructing a new, modern version of it by emblematically reutilizing the "Venice of the past, through a giant, anthropomorphized reproduction of the original . . . a new hybrid construction . . . a fetish shaped like a woman which in turn is represented through an anthropomorphic picture reutilizing an imaginative combination of ancient architectonic elements" (Ceccagnoli 2013, 179–80; Venezia del passato, mediante una riproduzione ingigantita e antropomorfizzata dell'originale . . . una nuova costruzione ibrida . . . un feticcio dalle fattezze di donna, che a sua volta

è rappresentata attraverso un'immagine antropomorfica che però riusa una fantasiosa combinazione di elementi architettonici del passato).

10. The Belgian multinational company Solvay owns the Rosignano plant. For a more detailed history of this process and an account of the related effects on the environment, see Baccaro and Musella (2013).

11. Nixon aptly observes that whenever "an official landscape—whether governmental, NGO, corporate, or some combination of those . . . is forcibly imposed on a vernacular one . . . it writes the land in a bureaucratic, externalizing, and extraction-driven manner that is often pitilessly instrumental" (2011, 17). It also happens that the "official landscape" is often welcomed with open arms by some of the less informed and needy inhabitants of the "vernacular one" because of its promise of future jobs, wealth, and socioeconomic improvement.

12. This promotional video is no longer displayed on the website of ENI's Historical Archive.

13. The slogan "Il cane a sei zampe è il miglior amico dell'uomo a quattro ruote" (The six-legged dog is the best friend of the four-wheeled man) is attributed to director Ettore Scola.

14. Most recently, Pasolini's *Petrolio* has inspired Adriatico's theatrical piece *Is, Is Oil* (2015), which proposes a dramatic reinterpretation of the text in the light of contemporary historical events (i.e., Isis terrorism, the situation in the Middle East).

15. Di Stefano's (2011) book collects a selection of the essays and short stories in *Il Gatto Selvatico* over the years.

16. Belloni writes that the magazine had three main goals: "cultivate and address the feeling of belonging to the company, valorize the historical aspect of the enterprise, and (especially) define and position ENI's institutional character and its business strategy within the national and international framework" (2014, 8). Interestingly, ENI's corporate publications after *Il Gatto Selvatico* were titled *Ecos* (1972–2002) and *ENI's Way Magazine* (2002).

17. See Menduni (1999).

18. As the *Oxford English Dictionary* specifies, *pedestrian* (adj.) means "lacking inspiration or excitement; dull."

19. Sullivan (2015) observes, "Like other new forms of the pastoral such as Terry Gifford's post-pastoral, the necropastoral in recent poetry, David Farrier's toxic pastoral, Greg Garrard's radical pastoral, the queer or gay pastoral, and the urban pastoral, the dark pastoral engages the traditional idyllic form's possible alternative to our current petrocultural consumer societies while attempting to avoid both sentimentalized views of static nature and also the naturalization of existing power structures."

20. Bontempelli's explanation of this feature is that

> its inventors and constructors . . . left a human particle of themselves in her. Official mechanics is ignorant about it and perhaps we shall never know where it is situated: it is a particle of that quality which makes man more of a man; that is to say his willpower.
>
> inventori e fabbricatori . . . le hanno senza accorgersene lasciato una particella umana di sé, la quale la meccanica ufficiale ignora e di cui forse non conosceremo mai la vera sede: una particella di quella qualità per la quale l'uomo è più uomo; cioè la capacità di volere. (1991, 51)

21. Pinkus observes not only that in their ads, "oil companies targeted drivers with images of carefree outings in the pristine countryside" but also that "the car as access to nature, represents, in essence, a conservative defense against modernism" (2012, 154).

22. As Kerényi writes, "In Arcadia there are mythological creatures whose exact features can be recognized in their primitive homeland, Tessaglia . . . centaurs . . . live in Tessaglia, in Arcadia and in the mountainous region in the Peloponnesous" (2014, 431).

23. In this regard, see also Pasolini (1965).

24. ENI's promotional video "Una raffineria in Lomellina" (A refinery in Lomellina) provides an antipodal depiction of an Italian

petrochemical plant as an almost heavenly site, ethereally and harmoniously in tune with the natural rhythms of rural Lombardy (ENI, n.d.). This commercial advertisement is constituted of a series of attractive images of the countryside, the farms, and the animals surrounding the refinery in different seasons. Located since 1963 in Sannazaro de Burgondi (in the province of Pavia), which has repeatedly been the site of accidents and explosions (the latest on July 2, 2016), the video never shows the refinery itself. Only in the last frame, in a well-staged night shot, are there some lights in the distance that could be those of a typical urban skyline.

25. See Sciascia and Ferrara (1964).

26. See Bonifazio for additional details about ENI's films, their educational role, and how they generally "excited and mobilized the emotions of audiences with lures of prosperity and happiness" (2014, 331).

27. Some of the darkest moments in recent Italian history are evoked by the titles of Ferrara's major works. Consider, for example, *100 giorni a Palermo* (1985), *Il caso Moro* (1986), *Giovanni Falcone* (1993), *I banchieri di Dio* (2002), and *Guido che sfidò le Brigate Rosse* (2007). At the beginning of his career, he additionally shot for ENI *Il Gigante di Ravenna* (1960), *Lucania* (1963), and once again on this particular Sicilian plant, *Il Gigante di Gela* (1964).

28. The poem is part of *Il falso e vero verde* (The false and real green), a collection of fourteen poems questioning the capitalist path Italy took soon after the end of World War II, whose title captures the situation of existential incertitude and environmental distress the poet addresses in several of his texts. See Quasimodo (1966).

29. Another extended literary quotation, from Elio Vittorini's short story "La Garibaldina," appears in a later section and introduces the narrator's remarks about the poverty of the ancient Gela vis-à-vis the apparent prosperity of the modern one.

30. The former happened in 405 B.C.E. and the latter in 1943. The narrator calls this tale "a Pascarelliana storia nostra," citing the work of poet and painter Cesare Pascarella (1858–1940).

31. See Susan Squier's remark: "Material conditions shape and reshape what we can put into words" (quoted in Alaimo 2010, 9).

32. At a later point, the plant is described as "this new organism that started to live and produce" (Sciascia and Ferrara 1964).

33. In her analysis of Daniele Vicari's documentary *My Country* (2006), Past observes, "Technological progress and myth are linked, and the miracle of petroleum discovery promises to elevate Sicily to the status of industrial giant" (2016, 376).

34. Reza Negarestani observes that "events are configured by the superconductivity of oil and global petro-dynamic currents to such an extent that the progression and emergence of events may be influenced more by petroleum than by time. If narrative development, the unfolding of events in a narration, implies the progression of chronological time, for contemporary planetary formations, history and its progression is determined by the influx and outflux of petroleum" (quoted in Pinkus 2016, 76).

35. Turri writes, "To be actors means to be 'internal' to the landscape, to be spectators means to be 'external' to it, a crucial distinction" (1998, 41; Essere attori significa stare "dentro" il paesaggio, essere spettatori significa starne "fuori," distinzione fondamentale).

36. The story of the wider sociocultural and environmental repercussions in certain areas of Southern Italy and of the controversial postunification policies that declined investing in and supporting the Bourbons' existing industrial heritage on the territory (i.e., iron and weapons factories in Campania and Calabria) is mostly still to be told.

37. Daniele Atzeni's docufiction *I morti di Alos* (The dead of Alos; 2011) is another short film set in Sardinia that, by focusing on a fictional environmental catastrophe, creatively addresses the costs of modernization on the island.

38. Mazzotta's *Oil* won the first prize at the 2009 Italian Environmental Film Festival (CinemAmbiente) in Turin.

39. Notably, a quote by director Giuseppe Ferrara praising *Oil* "for its militant rage with which it documents the criminality of an industrial system" appears at the beginning of the film.

40. On the representational limits of toxicity and the toxic image, see Schoonover (2013). As he concludes, "Even where the toxic resists precise depiction, its vestiges continue to haunt the image, registering something of the transformation of materiality that the toxic entails" (505).

41. A major intellectual figure in Sardinia and a proponent of the island's autonomous government but also, by necessity, a bilingual writer, Masala wrote for several journals and magazines, addressing what he perceived as the neocolonial imposition of standard Italian language over the traditional, regional dialect. As he wrote in the essay "Fra storia e autobiografia" (Between history and autobiography), now appended to the novel,

> I was born in a village of peasants and shepherds between Goceano and Logudoro, in northern Sardinia and, during my childhood, I heard speaking and I spoke only in Sardinian. When I was in first grade, the teacher, a strict man always dressed in black, forbade me and my classmates to speak in the only language we knew and forced us to speak in Italian, the "language of the Motherland," as he called it. That's how it happened that from being lively and smart, we all became witless and sad.
>
> Sono nato in un villaggio di contadini e pastori fra Goceano e Logudoro, nella Sardegna settentrionale e, durante la mia infanzia, ho sentito parlare e ho parlato solo in lingua sarda: in prima elementare, il maestro, un uomo severo e sempre vestito di nero, ci proibì, a me e ai miei coetanei, di parlare nell'unica lingua che conoscevamo e ci obbligò a parlare in lingua italiana, la "lingua della Patria," ci disse. Fu così che, da vivaci e intelligenti che eravamo, diventammo, tutti, tonti e tristi. (2001, 121)

42. As the narrator states in this diary, "Don Adamo appears in first, second and third person.... He compensates for the lack of his own identity with the simultaneous presence of three persons: one, two-in-one and three-in-one, that is to say, like God" (Masala 2001, 86–87; Don Adamo vi compare in prima, in seconda e in terza persona.... Compensa la propria mancanza d'identità, con la contemporanea presenza di tre persone: uno, bino e trino, insomma, come Dio).

43. Lehtimäki's (2013) discussion of Barbara Kingsolver's novel *The Lacuna* has been a useful source of inspiration here. Masala states that "the language of oil and tar ... is that of the Northern Italian bourgeoisie that ended the Risorgimento by industrially colonizing the South, but convincing us that it *had unified the Motherland*" (Masala 2001, 121; il linguaggio del petrolio e del catrame ... [è] la lingua della borghesia italiana del Nord, che ha concluso il Risorgimento colonizzando industrialmente il Sud, ma convincendoci di *aver unificato la Patria*).

44. During this same episode, the narrator introduces the notion of *glottofagia*, "a ruthless disease that destroys the language of migrated workers, devoured by that of the employers" (Masala 2001, 70; un male spietato che distrugge la lingua dei lavoratori emigrati, divorata dalla lingua dei datori di lavoro).

45. Don Adamo reveals, "I imagined ... my sex ... was some kind of refinery torch blazing a cloud of toxic sperm, or, better, an alembic dripping poisonous spermatozoa.... Laughter, like sex, is a veracious means of communication" (Masala 2001, 80, 83; Immaginai ... un sesso ... una specie di fiaccola di raffineria, vampante una nube di sperma tossico o, meglio, un alambicco gocciolante spermatozoi velenosi.... Il riso, come il sesso, è un verace mezzo di comunicazione). To further stress the relationship between the exterior and the interior, we may recall that the noun *Adamo* derives from the matter (*adamah*: "earth" or "ground" in Hebrew) from which he is made.

46. At the conclusion of the novel, Don Adamo states, "Therefore, don't even think

about knowing the epilogue of this story.... I too don't know how it will end. The beauty of life is that, differently from books, one can never write the word 'end'" (Masala 2001, 111; Perciò, non fatevi venire l'idea di conoscere l'epilogo di questa storia.... Nemmeno io so come andrà a finire. Il bello della vita è che, a differenza dei libri, non si può mai scrivere la parola "fine").

47. Don Adamo raises the possibility (only to immediately deny it) that he might have thrown Eva down the bell tower.

48. The dehumanizing and animalizing consequences of Sarroch's industrialization practices reappear even more clearly in the episode of Don Adamo's surreal encounter with one of his emigrated parishioners (Pietro Cocoi), whose new job is to impersonate a chimpanzee in a German circus (Masala 2001, 83–85).

49. Referring back to the beginning of this chapter, we may consider the feminized "desiring oil" that desperately wishes to see the light in Goretti's futurist poem as similarly "generative" in Barad's (2007) sense.

Chapter 3

1. The term *geocoreography* evokes Coole and Frost's notion of "choreographies of becoming," with which they indicate the relational activity and meshwork of "cosmic forces ... objects ... bodies ... subjectivities" and, in turn, an ontology in which "there is no definitive break between sentient and non-sentient entities or between material and spiritual phenomena" (2010, 10).

2. For an overall perspective and more information on the substantial environmental damages caused by current practices of marble extraction, see Mistiaen and Briganti (2015) and Stella (2015).

3. Barad explains that "the neologism 'intra-action' signifies *the mutual constitution of entangled agencies* ... in contrast to the usual 'interaction,' which assumes that there are separate individual agencies that precede their interaction, the notion of 'intra-action' recognizes that distinct agencies do not precede, but rather emerge through, their intra-action" (2007, 33).

4. Lopez writes that "to really come to an understanding of a specific American geography, requires not only time but a kind of local expertise, an intimacy with place few of us ever develop.... It resides with men and women more or less sworn to a place, who abide there, who have a feel for the soil and history, for the turn of leaves and night sounds" (1989, 53). His words could easily be applied to the particular Italian geography under examination and to the writers considered here.

5. Although I do not discuss his work, the name of another writer from this area, Sirio Giannini (1925–1960), could be added to this list. In both his major novels, *La valle bianca* (The white valley; 1958) and *Da dove nasce il fiume* (Where the river is born; 1971), the disintegrating social and environmental, human and nonhuman landscape of the Apuan Alps provides the background of the narration. As one of the characters in the second novel states, "Everything here crumbles, and even the families who continue to live here are like the latest stones in a landslide: they linger for a while hanging over the cliff, then down they go, joining the rest.... It feels like a body where all the blood in the smaller veins withdraws and goes to the larger ones, and never goes back.... Not even the stones in a landslide go up again to the top of the mountain from which they departed. If anything, they roll further away. They disperse" (1971, 23–24).

6. Together with artist Lorenzo Viani (1882–1936), Ceccardi, who was born in Genoa (Liguria), is the understudied bohemian bard of Lunigiana and the Apuane. Partially influenced by Giosuè Carducci and Gabriele D'Annunzio, scholars have praised his landscape-centered poetry and consider him the forefather of the "Ligurian line" of poetry later made famous by figures such as Camillo Sbarbaro and Eugenio Montale, who won the 1975 Nobel Prize in Literature.

7. Rigell usefully reminds us that

> a trade war affecting the main components of Sicily's economy (fruit, wine, and

sulfur) led the Sicilian elite to institute higher rents and taxes, transferring the economic hardship to the working class. Fasci, meaning "bundle" in Italian, were associations of laborers that advocated for workers' rights and constituted Sicily's first social movement.... Though the Fasci had expressed their commitment to non-violence, activism during the latter part of 1893 escalated beyond the control of the Fasci Central Committee.... In December, Crispi suppressed demonstrations with violence, resulting in 92 deaths. Fasci leadership disagreed on what approach to take if Crispi attempted to dissolve their organization. Though the majority of members recognized the imbalance of power, advocating for the use of nonviolence, some advocated for an insurrection. On January 3, 1894, expressing fear of the secession of Sicily from Italy, Crispi declared a state of siege, dispatching 40,000 troops to the island. Acting under martial law, he authorized the arrest of the members of the Fasci's Central Committee and ordered the dissolution of all Fasci. (2013)

8. About the link between Ceccardi's enlightened regionalism and his protoenvironmentalism, Pertici observes that "Ceccardo's regionalism ... does not contain elements of separatism and municipalism; it rather aims at recuperating local identities especially in opposition to the official policy of the Italian monarchy which appeared to him as a rootless entity within the real nation. One may thus better understand the real sense of his struggle to defend the environment and the artistic patrimony ... in a period when the first wave of the country's industrialization, the intense transformation of major cities, and the creation of infrastructures were deeply modifying and risked irreversibly altering urban and rural Italian landscapes" (2010, 31).

9. If it were not set in Sussex, Arthur Conan Doyle's 1928 short story "When the World Screamed" could easily have been set in the Apuan Alps, with the protagonist, the slightly mad scientist Professor Challenger, wishing to prove that deep underground, there is a geological layer that pulsates like a living organism.

10. It is worth noting that in this same period, the organic link between topology and human biology was being investigated in Friedrich Ratzel's *Politische Geographie*, a text introduced among the members of the Florentine School of geography thanks to a partial Italian translation by geographer and sociologist Cesare Battisti (1875–1916). Ratzel's "bio-geography," for example, stated that "the diffusion of mankind and its works on the Earth's surface has all the characteristics of a body" (quoted in Battisti 1988, 172; la diffusione degli uomini e delle loro opere sulla superficie terrestre porta tutte le caratteristiche di un corpo). This position was often viewed as a precursor of Nazism's notion of *lebensraum*.

11. Among marble's material effects on human bodies, one should also include the workers' alcoholism, which Ceccardi views (and justifies) as an environmentally related condition, directly linked with the place and the job they perform:

> True, those men, those quarrymen who today are going up rocks where goats can barely climb and tomorrow fall dashed to pieces are used to drinking on Saturdays and Sundays. Sometimes they also get drunk. But it is their hard life that asks for it. They need to invigorate, to forget even for just two or three hours, their youth wasted in the sun, their eyes bleeding because of the white reverberations.

> È vero quegli uomini, quei cavatori che oggi s'arrampicano per le roccie, dove appena salgono le capre e domani ne precipitano sfracellati, al sabato sera, alla domenica hanno l'uso del bere. Qualche volta s'ubriacano anche. Ma è la loro vita faticosa che lo richiede. Hanno bisogno di rinvigorirsi, hanno bisogno di obliare fosse pure per due o tre ore, la giovinezza sciupata al sole, la carne arsa, gli occhi sanguinanti pei bianchi riverberi. (1894, 10–11)

12. Locatelli and Anderson (2017) echo my point as they observe, "Follow Italy's marble,

and you follow the major movements of global wealth in human history, from ancient Rome to Victorian London to 20th-century New York."

13. Other relevant figures associated with this group in which Ceccardi was "the General" include Lorenzo Viani ("il Grande Aiutante"), Luigi Salvatori ("il Grande Cancelliere"), Giuseppe Ungaretti ("il Console d'Egitto"), and Enrico Pea ("il Sacerdote degli scongiuri"). What united them was, first, their restlessness and intellectual curiosity; second, their independence of judgment, anticonformism, and antibourgeois stance; and third, their republican faith and deep hostility toward the Savoia monarchy (see Bucciarelli, Ciccuto, and Serafini 2010, 9–10, 71).

14. Ungaretti's description of Pea's writing in a 1916 letter to Giovanni Papini poetically captures the interpenetration of the natural and the cultural, the human and the nonhuman:

> Sometimes he surprises you with moments so dense, so appropriate, so violent, so full of sky, with a humanity carved in a word still moist with soil and glittering with morning-dew, like a grass coming out to laugh in the sun in a beautiful morning: something that only Giotto and few others in this world have been able to do.
>
> Ha dei momenti che ti sorprendono per densità, proprietà, violenza, vastità di azzurro, per un'umanità intagliata in una parola tutt'ancora umida di terra, e brillante di rugiada, come un'erba spuntata a ridere nel sole, una mattina bella: come solamente Giotto e chi sa chi altro nel mondo hanno saputo fare (quoted in Appicciafuoco 2009, 5).

The following verses from a poem Pea dedicated to his friend, painter Giuseppe Viner (written on the back cover of Viner's copy of *Le fole*, Pea's first book of poetry), mirror Ungaretti's observation regarding Pea's words' closeness to the soil: "We have picked flowers underground, / to make a garland of joy with them, / for the creatures who are thirsty for love, / for the creatures who are thirsty for tears" (Pea 1910; Abbiamo colto fiori sotto terra, / per farne poi ghirlanda di letizia, / per le creature che han sete d'amore, / per le creature che han sete di pianto). Pea describes his book of poetry as "this tussock / that is ours, we picked it underground" (questo ciuffo d'erba / ch'è nostro, che l'abbiam colto sottoterra).

15. While in Alexandria, Pea created his own enterprise, called the Baracca rossa (Red shack), which was "the gathering place for an anarchist group and a reference point for any number of political refugees from all over Europe and the Mediterranean, as well as serving as the classroom for the 'Peoples' University' [Università popolare] [he also] founded" (Re 2003, 180).

16. In one of his radio speeches, Pound (1941) stated that Pea "writes like a man who could make a good piece of mahogany furniture."

17. As Oppermann observes, quoting Jeffrey J. Cohen and Lowell Duckert, "Ecomaterialism 'asks us to hear the howls of heterogeneous life forms—everywhere and from every thing . . . [it] compels us to think of our own existence as interstitial beings'" (2018, 121).

18. These include, for example, Scipio Slataper, Giovanni Boine, and Piero Jahier.

19. As Ben-Ghiat writes, "The Strapaese movement offered a way to build a national culture on autochthonous models" (2001, 26–27). Like others in Europe who felt Nietzsche's influence, several Italian modernist artists in these years (i.e., Giovanni Papini, Ardengo Soffici, Giovanni Boine, Piero Jahier, Clemente Rebora) thought that a return to nature, to the countryside, was necessary to achieve a new sense of rejuvenation and authenticity in one's life.

20. Interestingly, in the context of the anarcho-syndicalist theories of personal liberty and self-sufficiency, there were also attempts to overcome individualism in order to achieve solidarity with others (see Guarnieri 1979, xx).

21. After all, a vitalist understanding of reality informs Pirandello's aesthetic thought. As Perniola reminds us, "He ends up separating and juxtaposing the two classical notions

of life: life as *bios* (relating to the ethical and social dimensions of human behavior) and life as *zoé* (the naturalistic force that joins vegetables, animals and humans).... Pirandello believes that *bios* is the place of conceptual deceits, artistic forms, and social masks, and that only the absence of conceptuality of *vita nuda* [bare life] may reveal the deepest reality of human condition" (2003).

22. See, in particular, Somigli's analyses of the novel *One, No One, and One Hundred Thousand* and the short stories he discusses, "Canta l'epistola" (The epistle sings; 1911) and "Quand'ero matto" (When I was crazy; 1902), which complement my observations here. Somigli claims that in Pirandello, "over time, even nature loses its power to recompose the fragmented self in a post-death-of-God world" (2016, xvii). In the first story, Tommasino Unzio's "individual life has become so intertwined with that of one individual blade of grass that its fate is also his" (138). Tommasino reflects on humans' encounter with nature in passages like this: "To lose consciousness of being, just like a stone, like a plant; not to remember even one's own name; to live for the sake of living, without awareness of being alive, like animals, like plants; without attachments, or desires, or memories, or thoughts, without anything to give meaning and value to one's life" (137). In the second story, Somigli observes, "Fausto's self is not only open to others, but also to the nature around him.... His initial perception of the fundamental equality and reality of all beings which prevented him from putting his self-interest ahead of that of others translates, in the moment of communion with nature, into the intuition of the 'unity of Being.'" As Fausto says, "I also entered the life of plants and, little by little, from the pebble, from the blade of grass, I arose, gathering and feeling within me the life of everything, until I almost felt that I became the world, that the trees were my limbs, the earth my body, and the rivers my veins, and the air my soul; and I would go on for a while like this, in ecstasy and permeated by this divine vision" (quoted in Somigli 2016, 139).

23. As a curiosity, in the movie *Gli orizzonti del sole* (The horizons of the sun; 1955), directed by Giovanni Paolucci, Pea fittingly played the role of an old naturalist living on an island in Italy's Lake Maggiore who teaches a young man how to see the world around him.

24. All these factories were eventually closed, but as Rovelli's and Milani's books argue, the toxic effects and consequences of their presence on the territory and its population are still emerging today.

25. Milani's stance here recalls Pasolini (1975) and, more recently, the "facts-based fictions" by Saviano. The latter clarified his predilection for and the hybrid nature of this kind of speculative writing in an article in the newspaper *La Repubblica*. Here he praised the Nobel Committee for awarding the 2015 prize in literature to a nonfiction writer (Svetlana Alexievich). As Saviano puts it,

> Nonfiction ... is a literary genre whose goal is not a piece of news, but rather the story of the truth. The nonfiction writer prepares to work on a documentable truth, but s/he faces it with the freedom of poetry. S/he does not create the news, s/he uses them.... Adding reality to the novel, subtracting coldness from the news, are the only existing ways to bring "sensitive" topics to the readers' attention.

> La non fiction ... è un genere letterario che non ha come obiettivo la notizia, ma ha come fine il racconto della verità. Lo scrittore di narrativa non fiction si appresta a lavorare su una verità documentabile ma la affronta con la libertà della poesia. Non crea la cronaca, la usa.... Aggiungere realtà al romanzo, sottrarre freddezza alla cronaca, sono l'unica strada che esiste per portare argomenti "sensibili" all'attenzione del lettore. (2015)

Interestingly, in a relatable comment, Nixon observes that "the days when professors viewed the novel and poetry as the imagination's soaring peaks and nonfiction as the valley of the shadow of death may be over, but nonfiction is still widely treated in literature departments as at best a subsidiary form" (2011, 288).

26. A whole chapter of the book is dedicated to the story of an old partisan (Gardenio) who fought during World War II. He is the grandfather of Mauro, an environmental activist associated with the anarchic circles in Carrara, and one of Milani's main interlocutors.

27. Milani reveals marble's surprising ubiquity in our daily lives in the following passage:

> In the last two decades, the marble debris, the processing waste that fed . . . the mining dumps in the Apuan Alps became a "transversal commodity." Debris was pulverized in calcium carbonate and then used to produce plastics, rubber, tires, insulation, glues, paper, chemical and pharmaceutical products, cosmetics and building materials. Just to eliminate the sulfur oxides emitted by a coal-fueled, 1,000 Megawatt electric plant, you need 50,000 cubic meters [of pulverized marble debris] each year; 1,500 tons per year go into the toothpaste sold in Italy; and an undefined amount ends up in fodder and food . . . in the early nineties . . . the era of "Milan is for drinking" [the slogan of a popular ad for an alcoholic digestive] ended, and the one of the "Apuan Alps are for eating" began.
>
> I detriti di marmo, gli scarti di lavorazione che alimentavano . . . le discariche minerarie delle Alpi Apuane erano diventati, negli ultimi due decenni, "una categoria merceologica trasversale." Il detrito veniva polverizzato in carbonato di calcio e così impiegato per la produzione di plastiche, gomme, pneumatici, isolanti, vernici, colle, carta, prodotti chimici, farmaceutici, cosmetici e nell'edilizia. Solo per l'abbattimento degli ossidi di zolfo nelle emissioni di una centrale elettrica a carbone da 1.000 Megawatt ne servono 50.000 metri cubi all'anno; 1.500 tonnellate all'anno per il dentifricio venduto in Italia; poi una quantità indefinite nei mangimi e negli alimenti . . . nei primi anni novanta . . . si chiudeva l'era della "Milano da bere" e si apriva quella delle "Apuane da mangiare." (2015, 62–63)

Chapter 4

1. Actors Marcello Mastroianni, Alberto Sordi, and Paolo Panelli and screenwriter Suso Cecchi D'Amico all owned villas and used to vacation here.

2. I use the term *progressive* to emphasize how Tuscany's environmental efforts and policies are presented by the media (rightly so, in many cases) as particularly advanced if compared to those of other, euphemistically speaking, "historically less environmentally engaged" Italian regions (i.e., Campania or Calabria). For example, in 2019, the Blue Flag Program (Programma Bandiera Blu), which assigns a sought-after ecolabel to coastal localities that promote a sustainable management of the environment, listed several beaches: Livorno (Cala del Miramare, Rogiolo, Quercianella), Rosignano Marittimo (Castiglioncello, Vada), Bibbona (Marina di Bibbona), Castagneto Carducci, Cecina, Piombino (Parco naturale della Sterpaia), and San Vincenzo. See FEE (2019). This, of course, does not mean that the situation in Tuscany is ideal.

3. Interestingly, in his poem "In riva al mare" (On the shore of the sea), Carducci (1942, 31)—without any clear environmental concern in mind—already mentions the Tyrrhenian Sea's "dirty foaming waters" (sucide schiume) and its "unclean prey" (immonde prede) hunted by an enigmatic "cetacean" (cetaceo).

4. See Maltagliati and Barbaro (2012).

5. Bertrams, Coupain, and Homburg (2013) provide the reader with everything one wishes to know about Solvay and its praiseworthy initiatives and philanthropic corporate policies—with the exception of the history of the effects that this "family firm" has had for the past 150 years on the environment and in Rosignano Solvay and its nearby coastal areas as well. Of course, in typical fashion, since the Solvay plant plays a large role in the local economy, many residents

prefer to conveniently forget about pollution issues and, by either necessity or ignorance (or both), focus on the jobs it creates.

6. The Lucchini steel plant in Piombino is (still) in trouble. At the time of this writing, its workers are organizing strikes to protest the sale of the factory to an Indian steel-making company.

7. Silvia Avallone was born in Biella in 1984. After graduating with a degree in philosophy from the University of Bologna, she collaborated with *Il Corriere della Sera* and *Vanity Fair*. Her poetry and short fiction appeared in *Granta* and *Nuovi Argomenti*. The English translation of her *Acciaio* came out in 2012 with the title *Swimming to Elba*, which seems to target the same kind of Anglo-American reader who loves Frances Mayes's *Bella Tuscany* series of books (if so, these readers may not find exactly what they expect). A cinematic adaptation by Stefano Mordini appeared in 2012.

8. Slack and Wise observe how the machine has supplanted nature with "almost religious like reverence" (2005, 18).

9. Iovino and Oppermann point out that "apparently, the very dynamics of environmental degradation lie not only in our social and economic practices and imperialist attitudes to nature, but in the structures of the discursive formations that have led to such destructive mechanisms in the first place" (2012, 464). In this light, the comparison the novel makes between a cascade of melted steel and one of water conveys the former's apparent naturalness and therefore diminishes (or eliminates) the potential environmental concerns associated with it.

10. The following underlines the intermingling between human beings and industry: "Everyone knows, everyone takes it for granted that inside the Lucchini plant, deep down in the bowels of the place, the flesh of human legs, arms, and heads is stirring" (Avallone 2012, 73; Uno lo sa, lo dà per scontato, che dentro la Lucchini, nelle viscere, si muove la carne di gambe, braccia, teste umane).

11. Regarding "secretions," the fact that the narrator later refers to the plant as "the steel jungle, the incessant screeching, roaring, the ejaculations of the mill" (Avallone 2012, 291; la giungla d'acciaio, lo stridore continuo, ruggiti, eiaculazioni di impianti) reveals again the confusion between technology and biology.

12. This image makes me think of David del Principe's ecogothic approach, which also encompasses questions of industrialization in the nineteenth century.

13. I am borrowing the concept of "necroregion" from Iovino (2012b) to indicate an extension of space (such as the cane-brake but also the whole area of the Lucchini factory) that shows signs of ecological and cultural abandon.

14. In the section "Thing-Power I: Debris," Bennett describes a (rather similar) encounter she had "on a sunny Tuesday morning . . . in the grate over the storm drain to the Chesapeake Bay" and writes, "Glove, pollen, rat, cap, stick. . . . I caught a glimpse of an energetic vitality inside each of these things, things that I generally conceived as inert" (2010, 4–5).

15. *Ilva* is the Latin word for Elba. The mineral ilvaite, a silicate of calcium and iron, and the ILVA steel corporation were named after it.

16. Taranto's ILVA is responsible for most of the dioxin output in Italy. As Seger observes, its "residents have long faced startlingly excessive mortality rates compared to the general Italian population, and talk of tumors is a daily occurrence in the city's neighborhoods and bars" (2018, 185).

17. Gregson, Watkins, and Calestani note,

> Asbestos is the generic name for a group of silicate minerals with a long thin fibrous structure. There are various forms of the mineral, with chrysotile (white), amosite (brown) and crocidolite (blue) being the most commonly used in industrial and domestic applications in the twentieth century. Chrysotile accounts for greater than 90 percent of asbestos currently being used commercially. It is the most flexible of the asbestos fibers, in that its curly fibers withstand the fiercest heat yet are so soft and flexible that they can

be spun and woven as easily as cotton. . . . Currently it has over 3,000 patented uses. (2010, 1065n1, 1066)

While its extraction and commercialization were banned in all EU countries in 1999 and in Australia in 2003, it is still widely used, legally, in India, China, and yes, even in the United States. Gregson, Watkins, and Calestani continue, "In 1989, the EPA [Environmental Protection Agency] issued a final rule under Section 6 of Toxic Substances Control Act (TSCA) banning most asbestos-containing products. However, in 1991, this rule was vacated and remanded by the Fifth Circuit Court of Appeals. As a result, most of the original ban on the manufacture, importation, processing, or distribution in commerce for the majority of the asbestos-containing products originally covered in the 1989 final rule was overturned" (1065). Estimates from the World Health Organization (WHO) report that 125 million people are exposed to asbestos annually.

18. The relative insignificance and marginal agency of human workers in this place (also from a labor perspective), as well as the interpenetration between the natural and the cultural, is conveyed by Calvino in his initial description: "The mountain sloped piece by piece into the factory's mills, and was spitted out again to form enormous heaps of slag, which constituted a new, still shapeless, matt gray mountainous system" (1954, 192; La montagna scendeva pezzo a pezzo nei frantoi della fabbrica, e veniva risputata in enormi cumuli di scorie, a formare un nuovo, ancora informe sistema montuoso grigio opaco). At the same time, this image tells us much about industrialization processes in the Italian Alps and, in general, about how various risks tend to arise whenever industrial capitalism encloses nature in the factory system and transforms it into a property (see Barca 2010a, 4).

19. Prunetti fully captures the insidious agency of asbestos fibers when he describes them at the very beginning of his book as

> small crystal-like darts. They are invisible arrows capable of going down the esophagus, diving into the lungs and sticking to the pleura for twenty, thirty, even forty years, causing a badly cicatrized wound that the organism cannot vanquish, and that initiates a process of cellular degeneration. A tumor.
>
> piccoli dardi cristallini. Saette invisibili capaci di scendere lungo l'esofago, di calarsi nei polmoni e di rimanere attaccate alla pleura per venti, tranta, anche quarant'anni, producendo una ferita mal cicatrizzata che l'organismo non riesce a debellare e che avvia un processo di degenerazione cellulare. Un tumore. (2012, 16).

20. As Wu Ming 1 states, *Amianto* is one more example of those "'Uno, Unidentified Narrative Objects'" we previously encountered: "[It is] not only an investigation, a biography, a novel, an essay, but it has small doses of each of these components" (quoted in Prunetti 2012, 176; Non è inchiesta, non è biografia, non è romanzo, non è saggio, ma ha ognuna di queste componenti in piccole dosi).

21. Jansen notes that by mythologizing Renato, Prunetti both prevents his victimization and "transposes his father's private memory into a collective memory of workers' resistance against capitalism" (2017, 22).

22. Here I am thinking of Branch's (2014) insights.

23. Eternit had other plants both in Italy and in other parts of Europe and South America.

24. I share Lindgren and Phillips's observations about the unfairly unrecognized value of "the story-gathering and storytelling function of journalism . . . especially in areas of [asbestos] illness narrative" (2016, 162).

25. Reactive Citizens's goal is to create "investigative journalism in the public interest" and, in particular, to support "a new generation of . . . citizens able to perform civil monitoring, protect common good and share knowledge" (see Battaglia 2013). Battaglia's recent investigative documentary *La rivincita di Casale Monferrato* (Casale Monferrato's revenge; 2018) was crowdfunded over four years through the Reactive Citizens online platform. It expands on earlier films, such

as Niccolò Bruna and Andrea Prandstaller's *Polvere, il grande processo dell'amianto* (Dust, the great asbestos trial; 2011), by shifting its focus on the hearings of the trial against Eternit and its (lack of) corporate responsibility and on a tragic individual story (that of Luisa Minazzi, a local activist who got sick with mesothelioma) to a more positive outlook. Specifically, it illustrates the collective fight of Casale's citizens to reclaim their city by remembering their past and looking forward to a brighter future by cultivating "justice, reclamation [and] research." The episode "Amianto: Le morti silenziose" (Asbestos: The silent deaths) in the RAI 3 TV program *Blu notte* (February 13, 2012) is another example of quality investigative journalism and storytelling. In this episode, the host (writer Carlo Lucarelli) recounts the history of Eternit in Casale Monferrato by weaving together several reports of citizens, ex-workers, and their family members. See Lucarelli (2012).

26. Giuseppe "Pippo" Fava (1925–1984) was a Sicilian investigative journalist, playwright, novelist, and antimafia activist. He was killed because he consistently denounced the intertwinement of finance, politics, and drug traffic in Sicily and beyond.

27. Jansen clarifies that "the concept of 'transmediality' is used for 'phenomena, such as narrative itself, whose manifestation is not bound to a particular medium" (2017, 27n5).

28. This second section of *Malapolvere* invites a parallel with Armiero (2014b). In this latter volume, however, women actively write and reflect on the toxicity of their lives and land in Campania rather than telling their stories to a journalist who, in turn, puts them on paper.

29. Curino's (2012) monologue was originally broadcast live on RAI's Radiotre on November 22, 2012.

30. On the links of landscape practices with dominant forms of power and resistance, see Casid (2005).

31. Here I catch myself echoing Guattari, who wrote about the "erroneousness of portioning the real into a number of separate fields" and the mistake of regarding "action

on the psyche, the socius, and the environment as separate" (1989, 34).

Chapter 5

1. The affinities between *Clockwork* and Trevisan's (2010) vision emerge with particular force in the following passage:

> In this immense, consciousless, polycentric periphery everything is thought in pieces, and reduced over and over in pieces, just like its roads, its countryside et cetera; and, expectedly, the pieces are always smaller, and they are in danger of becoming so small that they cannot be fragmented any further.... [There is] a background noise that... constantly accompanies us, day and night... accepted as something "natural," so that there is no mutation in this fog, which is rather a dust.

> In questa grandissima periferia policentrica che non ha coscienza di sé, tutto è pensato a pezzi, e fatto e rifatto a pezzi, proprio come le sue strade e le sue campagne eccetera; e i pezzi, com'è ovvio, sono sempre più piccoli, e rischiano di diventare così piccoli da non permettere più di esser fatti ulteriormente a pezzi.... Un rumore di fondo che... ci accompagna costantemente, in ogni momento del giorno e della notte... accettato come qualcosa di "naturale," così che non c'è mutamento in questa nebbia, che è piuttosto una polvere. (Trevisan 2010, 17–20)

2. Magnaghi (2005) proposes, in addition to conventional indices of wealth and income, other criteria of social progress—including quality of life, social solidarity, and the development of noncommercial caring relations. Only when local communities value their local heritages, he argues, and build themselves on a basis of local economic self-government will they be able to resist the colonization and marginalization that globalization so often inflicts on them.

3. Cederna adds that "in the last three decades, we have submerged one-fifth of Italy (about six million hectares) under cement

and asphalt . . . within three or four generations, all of Italy will be . . . covered from North to South by a continuous and repellent crust made of buildings and roads" (1991, 10). On the topic of overbuilding, specifically in Lombardy, see also Petitto (2012).

4. Designed by Giuseppe Sacconi in 1855 and built in honor of Victor Emmanuel II, the first king of a unified Italy, the Vittoriano, as it is also known, is often considered conspicuous, pompous, and too large.

5. Wu Ming 2 (2016) chronicles a walking trip along the TAV between Bologna and Milan. The protagonist's conclusive reflections are dedicated to Expo 2015.

6. As an example of this erasure and "total removal of one of the pillars of [Milan's] working-class history" (Wu Ming 2 2016, 263; totale rimozione di un pilastro della sua storia operaia), Wu Ming 2 mentions the Alfa Romeo plant that, as he puts it, was "dissolved in acid as if it were a cumbersome cadaver" (263; sciolto nell'acido come un cadavere ingombrante).

7. These are the groups listed on the Environmental Justice Atlas (EJAtlas): Comitato NoExpo, Attitude No Expo, Italia Nostra, Milan's Polytechnic University, Rete metropolitana contro debito, cemento e precarietà, La Terra Trema, Milano in Movimento, and M5S. See EJAtlas (2015).

8. The terms *impermanent* and *disposable* recall what Jackson describes as the emergence, "particularly over the last half century, [of] a vast number of structures designed and built to last for a period measured in a few years if not in months" (1984, 111). However, the adjectives do not capture the far-from-ephemeral extent of Expo's impact on the territory.

9. Trevisan (2010) uses this expression to describe "'a fragmentation, a parceling, a causal succession of territorial elements" (quoted in Valdemarca 2014, 86; una frantumazione, parcellizzazione, successione causale di elementi del territorio) and, as Valdemarca explains, "the total abandonment of the notion of center, which is substituted by an indeterminacy that characterizes every level—urbanistic, economic, ecological and subjective" (86; l'abbandono totale del concetto di centro per un'indeterminatezza che caratterizza ogni livello, sia quello urbanistico, paesaggistico, economico, ecologico e soggettivo).

10. For a complementary reading of Falco's collection that focuses also on its formal and stylistic features from a (neuro)narratological perspective, see Tomasi (2014). Giorgio Falco was born in Abbiategrasso (Lombardy) in 1967. His name may be added to the list of northern writers—from the canonical (Gianni Celati and Andrea Zanzotto) to newer ones (Vitaliano Trevisan, Massimiliano Santarossa, and Romolo Bugaro)—who have been variously referring in their novels to some of the drastic environmental transformations in the Italian North.

11. Wu Ming 2 (2017) provides a vivid description of this geographic area and its monstrosities.

12. On Seveso, see especially Seger (2017) and Centemeri (2010).

13. Tomasi observes that Falco offers "the portrait of a suburban formation that looks like a sort of *theme park* built by dull screenwriters . . . in which economic status becomes the only element able to define a role or to allow inclusion or exclusion" (2014, 101; il ritratto di una formazione suburbana che sembra una sorta di *theme park* costruito da sceneggiatori privi di fantasia . . . nel quale, unico elemento capace di definire un ruolo o di permettere l'inclusione o l'esclusione, diventa lo status economico).

14. On the logic that informs the choice of names for real estate developments, see Mauro Varotto's darkly entertaining observations in his chapter "Abitare tra le isole del Veneto centrale" (Living among the islands of central Veneto) in Vallerani and Varotto (2005, 79–82).

15. It is always important to remember that "human ideas, meanings and values shape and are shaped by . . . the environment out there" (Neimanis, Åsberg, and Hedrén 2015, 71–72).

16. I use *mind* in Gregory Bateson's sense of "a principle that is 'immanent' to all structures

and objects, be they natural or cultural" (quoted in Gersdorf and Mayer 2006, 15).

17. See, for example, the following description: "The flesh disperses in the five, ten, twenty meters ahead . . . it stains the car's exterior, or it is flattened until it becomes a camouflaged sign, where it is unthinkable to distinguish the head, the legs, the tail; in the end, it's only asphalt" (Falco 2009, 47; La carne si disperde nei cinque, dieci, venti metri avanti . . . macchia la lamiera, oppure resta appiattita fino a diventare segno, mimetizzato dove è impensabile decifrare la testa, le zampe, la coda, infine solo asfalto).

18. See Mitman, Armiero, and Emmett (2018).

19. The rest of this passage reads, "If it were not for the polluted air that, mixed with the smell of charcoal and grilled meat leftovers, stagnates over the young, green hedges, here [in Cortesforza] one feels like one is at the beach" (Falco 2009, 11; Se non fosse per l'aria inquinata che ristagna appena sopra il verde delle giovani siepi e si unisce all'aroma di carbonella e ai residui di carne, qui è come se fosse il mare).

20. Both this and the previous story invite one to think of Calvino (1991a).

21. On the narrative role of cars in these stories as "a further instrument to 'plastify' [in the sense of fictionalize, filter, mediate] reality," see Tomasi (2014, 102–4).

22. Ferroni defines *autofiction* as a textual situation in which the first-person narrator "is not exactly autobiographical, but s/he is also not entirely fictional, the 'I' coincides as a whole or in part with the real author's 'I' and may even take her/his name, even though s/he may add, at various degrees, fictional facts onto her/his personal events" (2010, 83). In the opening pages, Wu Ming 2 describes what his work is not: "Not a travel guide . . . not a novel . . . not a collection of short stories . . . not a diary, not an essay or a journalistic inquiry" (2016, 3; Sarebbe dunque difficile affermare che questa è una guida per escursionisti . . . un romanzo . . . una raccolta di novelle . . . un diario di bordo . . . un saggio, un reportage o un'inchiesta).

23. Sarchi's *Violazione* (which won the Paolo Volponi Opera prize) was followed by *L'amore normale* (Normal love; 2014) and *La notte ha la mia voce* (The night has my voice; 2017).

24. The expression is in Antonello (2009).

25. The epigraph, quoted in Sarchi (2012), reads, "NATURE: 'Thus the squirrel escapes the rattlesnake, until it ends up in its fangs on its own. I am the one you are escaping from.' / ISLANDER: 'Nature?' / NATURE: 'Precisely'" (NATURA: "Cosí fugge lo scoiattolo dal serpente a sonaglio, finché gli cade in gola da se medesimo. Io sono quella che tu fuggi." / ISLANDESE: "La Natura?" / NATURA: "Non altri").

26. The text suggests that Vanessa's handicap may be due to the fact that Primo and his wife, Genny, "had started to have sex too early [after her first pregnancy] without worrying about the consequences" (10; troppo presto avevano ripreso a fare sesso senza preoccuparsi delle conseguenze). Sadly, waste and substandard materials are also used in Primo's building projects.

27. Jon has a mostly sonic, animallike perception of language that for him eventually becomes an "undecipherable sound": "The sense [of nouns] completely disappeared and just their undecipherable sound was left" (Sarchi 2012, 15; Il senso spariva del tutto e rimaneva solo il suono indecifrabile).

28. Donati sensibly points out that Primo Draghi's character both "within the narrative economy of *Violazione* as well as in the real economy of the country . . . is not the sickness but the symptom, not the malfunctioning of an otherwise impeccable mechanism, but the proof that such a mechanism is *itself* defective" (2016, 168).

29. In her review of *Violazione*, Pugno provides a scathing portrait of Linda and Alberto as "apparently nice, but poisoned by the quintessentially Italian vice of being complacent in front of power" (2012, 11; apparentemente buoni, ma carsicamente inquinati dal vizio tutto italiano dell'accondiscendenza).

30. The following description conveys the sense of an extended nonplace: "Industrial containers, commercial centers and roads . . . connect pieces of that land which one cannot define as countryside, periphery or even less a city . . . the Dolomites

are becoming sand, because of their age more than anything else" (Sarchi 2012, 70; Container industriali, centri commerciali e strade... collegano pezzi di quella terra che non si riesce piú a definire campagna, ma nemmeno periferia, o tantomeno città... le Dolomiti si stanno riducendo a sabbia, per vecchiaia piú che altro).

31. The fragility of the land emerges in the following lines: "This is a land of gullies that, little by little, crumbles like sugar in water" (Sarchi 2012, 102; Questa è terra di calanchi, che un po' alla volta si sciolgono, come zucchero in acqua) and "That land was ugly, dark, treacherous, always ready to slide down and crack in the summer, and heavy as lead in the winter when it was wet" (143; Quella terra era brutta, scura, infida, sempre pronta a franare e creparsi d'estate, pesante come il piombo d'inverno quand'era bagnata).

32. Farina underlines from yet another angle the role and relevance of narratives as "informational tools" when he observes that "unlike economy, ecology is a science that studies natural machines and does not have operational tools able to modify [the functions and speed of the earth's gears], but only educational tools... to inform about the consequences of ecosystem change for human well-being, alerting people to present and future damage observed in a natural modified ecosystem" (2010, 2).

33. The notion of "Well-Being vs. Ill-Being Landscape" is Farina's (2010, 183).

34. From the Greek *apokaluptein*, "uncover, reveal": *apo-*, "un-," and *kaluptein*, "to cover."

35. As the ISPRA report states,

> The classical [1960s] polarities... have been progressively losing clarity, offering instead an opaque scenery [constituted by] a latent and widespread urbanization which invades coastal areas and occasionally swells towards the interior. [This happens] in Tuscany as well as in Calabria, in Abruzzo and in Basilicata, covering more than 7% of the national surface... most of the growth of cementified areas is no longer concentrated around urban areas, but is dispersed in large agricultural spaces, creating voids and almost never filling them up.

> Le classiche polarità... perdono progressivamente nettezza, restituendo uno scenario opaco, di urbanizzazione latente e diffusa, che invade le aree di costa e si dilata a tratti verso l'interno, in Toscana come in Calabria, in Abruzzo come in Basilicata, estendendosi per oltre il 7% della superficie nazionale... la crescita maggiore delle aree cementificate non è più concentrata intorno alle aree urbane, ma si disperde nei grandi spazi agricoli, creando vuoti e quasi mai colmandoli. (2017, 159)

36. Lorenzo de' Medici notoriously set his *poemetto rusticano* "Nencia da Barberino" (ca. 1469–73) in Mugello and parodically emphasized its rurality and the unsophistication and backwardness of its inhabitants with respect to the nearby city of Florence.

37. The A1, a.k.a. Autostrada del Sole (Sun Motorway), is Italy's major highway. As Menduni reminds us, it was inaugurated in 1964, originally stretched 755 kilometers from Milan to Naples, and was meant to symbolize the "solidarity between North and South and national unity" (1999, 70–71).

38. She aptly calls them "migrants of the great works" (Baldanzi 2011, 112; migranti delle grandi opere), thus individuating a new, contemporary form of internal South-North migration that mirrors the one that took place in Italy in the 1960s.

39. Baldanzi writes that in Norbert Elias's *Society of Individuals* (1991), Elias's understanding of the ontological compenetration between the individual and the social dimensions provides an important methodological base for her research. Accordingly, the Mugello and the workers together form "a *figuration*, a specific form derived from their relationship, a structure in tension and continually moving" (Baldanzi 2011, 32n5; una *figurazione*, una forma specifica scaturita dal loro rapporto, una struttura in tensione e in continuo movimento). I am intrigued by the apparent affinity of this approach with some of the holistic positions

of ecomaterialism—particularly the idea that any environment is constituted by the interaction of material subjects over time and space.

40. The tension in *Statale 18* between Cresswell's two forms of metaphysics (sedentary and nomadic) might be synthesized in the narrator identifying himself as "a pedestrian who uses the car" (Minervino 2010, 210; un viandante che usa l'automobile).

41. See, for example, the narrator's definition of SS18 as a "kind of Via Emilia situated in Calabria" (Minervino 2010, 58; questa specie di Via Emilia delle Calabrie), inviting us to associate these two roads and introducing a disorientation that erases distances and the stereotypical differences between Northern and Southern Italy. The devastating spatial homologation he describes flattens geographies and reduces every place to a nonplace.

42. Minervino nostalgically writes of ancient "Mediterranean suburbs where flourished terraced gardens, orchards, orange, lemon and mulberry groves, olive and fig trees, and vineyards" (2010, 220; sobborghi mediterranei dove vegetavano orti terrazzati, verzieri, aranceti, limoneti, gelseti, piante di olivi e olivastri, vigne e ficare), which have given way to gray urban peripheries.

43. The following depiction of SS18 illustrates this point (while also evoking the image of a tunnel): "A hybrid with no end, a cylindric container, a blast furnace in which a portion of the planetary chaos melts down" (Minervino 2010, 59; Un ibrido senza fine, un contenitore cilindrico, un altoforno in cui si squaglia una porzione del caos planetario).

44. As it retraces the events linked to the criminal sinking of various ships containing toxic waste off the Tyrrhenian coast of Calabria, the section entitled "La nave dei folli" (The ship of fools) only apparently detours from SS18, since it underlines how local mafias maintain their power by occupying space everywhere they can and explores the extent of their impact on this highly permeable terrestrial and marine habitat.

45. My conclusive reflections are bolstered by the fact that the image of tuna fish resurfaces in the narrator's mind at the end of the book: "I am thinking once again of the tuna fish, of the magnificent tuna who are swimming fast, attracted by the eye of the moon" (Minervino 2010, 234; Penso di nuovo ai tonni, ai magnifici tonni che nuotano veloci attratti dall'occhio della luna).

Epilogue

1. The notion of a wider sense of agency beyond the human and humankind's intentionality is conveyed by the following remark: "Willpower exists, but it functions in response to many factors including linguistic, cultural, biological, chemical, and ecological ones" (Dürbeck, Schaumann, and Sullivan 2015, 122).

2. This said, I share Breu's position when he writes, "It is crucial that posthumanism and the materialist turn not participate in a devaluation of human life or a rejection of human concerns, even as we rightly expand our ethical, political, and economic concerns beyond the circle of the human" (2014, 193).

References

Adamson, Joni, William A. Gleason, and David N. Pellow. 2016. *Keywords for Environmental Studies*. New York: New York University Press.

Adorno, Salvatore. 2010. "Petrochemical Modernity in Sicily." In *Nature and History in Modern Italy*, edited by Marco Armiero and Marcus Hall, 180–94. Athens: Ohio University Press.

Adriatico, Andrea. 2015. *Is, Is Oil*. Andreaadriatico.it. Accessed April 2020. http://www.andreaadriatico.it/site/is-is-oil/.

Agenzia Nazionale Turismo. n.d. "Italia." Italia.it. Accessed April 2020. http://www.italia.it/en/home.html.

Alaimo, Stacy. 2010. *Bodily Natures: Science, Environment, and the Material Self*. Bloomington: Indiana University Press.

———. 2016. *Exposed: Environmental Politics and Pleasures in Posthuman Times*. Minneapolis: University of Minnesota Press.

———. 2017. Introduction to *Gender: Macmillan Interdisciplinary Handbooks*, edited by Renée C. Hoogland, xiii–xviii. New York: Macmillan.

Alaimo, Stacy, and Susan Hekman. 2008. *Material Feminisms*. Bloomington: Indiana University Press.

———. 2015. "Emerging Models of Materiality in Feminist Theory." In *Ecocriticism: The Essential Reader*, edited by Ken Hiltner, 143–54. Abingdon: Routledge.

Alvaro, Corrado. 1995. *Itinerario italiano*. Milan: Bompiani.

Amberson, Deborah. 2018. "Temporalities of Stone in Scipio Slataper's *Il mio Carso*." *Italian Culture* 36 (1): 1–17.

Antonello, Pierpaolo. 2009. *Postmodern Impegno: Ethics and Commitment in Contemporary Italian Culture*. Bern: Peter Lang.

Appicciafuoco, Ilenia. 2009. "Il romanzo di Moscardino: Per riscoprire un autore di razza." Retididedalus.it, July 2009. http://www.retididedalus.it/Archivi/2009/luglio/LUOGO_COMUNE/pea/pea.pdf.

Armiero, Marco. 2011. *A Rugged Nation: Mountains and the Making of Modern Italy*. Cambridge: White Horse Press.

———. 2014a. "Introduction: Fascism and Nature." *Modern Italy* 19 (3): 241–45.

———. 2014b. *Teresa e le altre: Storie di donne nella terra dei fuochi*. Milan: Jaca Book.

Armiero, Marco, and Marcus Hall. 2010. *Nature and History in Modern Italy*. Athens: Ohio University Press.

Armiero, Marco, and Ilenia Iengo. 2017. "The Politicization of Ill Bodies in Campania, Italy." *Journal of Political Ecology* 24:44–58.

Arminio, Franco. 2011. *Terracarne: Viaggio nei paesi invisibili e nei paesi giganti del Sud Italia*. Milan: Mondadori.

Arrigoni, Fabrizio. 2009. *Cava: Architettura in "ars marmoris."* Florence: Florence University Press.

Asor Rosa, Alberto. 2014. "Le Alpi Apuane rischiano il disastro." *Il Tirreno*, June 25, 2014. https://iltirreno.gelocal.it/versilia/cronaca/2014/06/25/news/asor-rosa-le-alpi-apuane-rischiano-il-disastro-1.9487850.

Augé, Marc. 1995. *Non-places: Introduction to an Anthropology of Supermodernity*. Translated by John Howe. London: Verso.

Avallone, Silvia. 2012. *Swimming to Elba*. Translated by Antony Shugaar. New York: Penguin. Originally published as *Acciaio* (Milan: Rizzoli, 2010).

Baccaro, Andreina, and Antonio Musella. 2013. *Il paese dei veleni: Biocidio. Viaggio nell'Italia contaminata*. Rome: Round Robin Editrice.

Baldanzi, Simona. 2011. *Mugello sottosopra: Tute arancioni nei cantieri delle grandi opere*. Rome: Ediesse.

Balla, Giacomo, and Fortunato Depero. 1915. "Ricostruzione futurista dell'universo." Quellodiarte.com. Accessed April 2020. https://quellodiarte.com/documenti/ricostruzione-futurista-delluniverso/.

Banita, Georgiana. 2014. "From Isfahan to Ingolstadt: Bertolucci's *La via del petrolio* and the Global Culture of Neorealism." In *Oil Culture*, edited by Ross Barrett and Daniel Worden, 145–68. Minneapolis: University of Minnesota Press.

Barad, Karen. 2007. *Meeting the Universe Halfway: Quantum Physics and the Entanglement of Matter and Meaning*. Durham: Duke University Press.

Barca, Stefania. 2010a. *Enclosing Water: Nature and Political Economy in a Mediterranean Valley, 1796–1916*. Cambridge: White Horse Press.

Barrett, Ross, and Daniel Worden, eds. 2014. *Oil Culture*. Minneapolis: University of Minnesota Press.

Barron, Patrick. 2007. Introduction to *The Selected Poetry and Prose of Andrea Zanzotto*, edited and translated by Patrick Barron, 1–17. Chicago: University of Chicago Press.

Battaglia, Rosi. 2013. "Civic Journalism: Su ambiente, salute e legalità." Cittadinireattivi.it. Accessed April 2020. http://www.cittadinireattivi.it.

———. 2018. *La rivincita di Casale Monferrato*. Rosibattaglia.it, October 22, 2018. http://www

.rosybattaglia.it/tag/la-rivincita-di-casale-monferrato.

Battisti, Cesare. 1988. *Carteggi 1894–1916*. Edited by Vincenzo Calì. N.p.: Edizioni Temi.

Belloni, Eleonora. 2014. "L'ENI e il Terzo Mondo: L'anticolonialismo di Enrico Mattei nelle pagine de *Il Gatto Selvatico*." *Progressus: Rivista di storia* 1 (2): 2–14.

Benedetti, Carla. 1995. "Per una letteratura impura." In *A partire da Petrolio: Pasolini interroga la letteratura*, edited by Carla Benedetti and Maria Antonietta Grignani, 9–13. Ravenna: Longo.

Ben-Ghiat, Ruth. 2001. *Fascist Modernities: Italy, 1922–1945*. Berkeley: University of California Press.

Bennett, Jane. 2010. *Vibrant Matter: A Political Ecology of Things*. Durham: Duke University Press.

Benvegnù, Damiano. 2018. *Animals and Animality in Primo Levi's Work*. London: Palgrave McMillian.

Berman, Marshall. 1982. *All That Is Solid Melts into Air*. New York: Penguin.

Bertrams, Kenneth, Nicholas Coupain, and Ernst Homburg. 2013. *Solvay: History of a Multinational Family Firm*. Cambridge: Cambridge University Press.

Bevilacqua, Mirko. 1977. "'Le monoplane du pape': Il breviario estetico-politico di F. T. Marinetti." In *Marinetti futurista*, 257–65. Naples: Guida.

Bevilacqua, Piero. 2010. "The Distinctive Character of Italian Environmental History." In *Nature and History in Modern Italy*, edited by Marco Armiero and Marcus Hall, 15–32. Athens: Ohio University Press.

Blum, Cinzia Sartini. 1996. *The Other Modernism: F. T. Marinetti's Futurist Fiction of Power*. Berkeley: University of California Press.

Boeri, Stefano. 2011. *L'anticittà*. Rome: Editori Laterza.

Bond, Emma. 2016. "'Intoxicated Geographies': Sites of Refraction and Fragmentation in Scipio Slataper's *Il mio Carso* and Hermann Hesse's *Der Steppenwolf*." *Modern Language Review* 111 (1): 116.

Bonifazio, Paola. 2014. "United We Drill: ENI, Films, and the Culture of Work." *Annali d'Italianistica* 32:329–50.

Bontempelli, Massimo. 1991. "522": *Racconto di una giornata*. Rome: Lucarini.

Bozak, Nadia. 2012. *Cinematic Footprint: Lights, Camera, Natural Resources*. New Brunswick: Rutgers University Press.

Braidotti, Rosi. 1994. *Nomadic Subjects: Embodiments and Sexual Differences in Contemporary Feminist Theory*. New York: Columbia University Press.

———. 2013. *The Posthuman*. Cambridge: Polity Press.

Branch, Michael P. 2014. "Are You Serious? A Modest Proposal for Environmental Humor." In *The Oxford Handbook of Ecocriticism*, edited by Greg Garrard, 377–90. Oxford: Oxford University Press.

Breakthrough Institute. n.d. "The Breakthrough Institute." Accessed April 2020. https://thebreakthrough.org/.

Breu, Christopher. 2014. *Insistence of the Material: Literature in the Age of Biopolitics*. Minneapolis: University of Minnesota Press.

Brown, Bill. 2013. "Materialities of Modernism: Objects, Matter, Things." In *A Handbook of Modernism Studies*, edited by Jean-Michel Rabaté, 281–95. Somerset, NJ: John Wiley & Sons.

Bucciarelli, Stefano, Marcello Ciccuto, and Antonella Serafini, eds. 2010. *La Repubblica di Apua*. Florence: Maschietto Editore.

Calvino, Italo. 1954. "La fabbrica nella montagna." Academia.edu. Accessed April 2020. https://www.academia.edu/34564942/La_Fabbrica_nella_montagna_di_Italo_Calvino, 192–93.

———. 1991a. "La formica Argentina." In *Romanzi e racconti*, vol. 1, 445–82. Milan: Mondadori.

———. 1991b. *La speculazione edilizia*. Milan: Mondadori.

———. 1994. "Dialogo con una tartaruga." In *Romanzi e racconti*, vol. 3, 1155–58. Milan: Mondadori.

Campbell, Timothy. 2009. "Vital Matters: Sovereignty, Milieu, and the Animal in Futurism's Manifesto." *Annali d'Italianistica* 27:157–73.

Canavan, Gerry. 2014. "Retrofutures and Petrofutures: Oil, Scarcity, Limit." In *Oil Culture*, edited by Ross Barrett and Daniel Worden, 331–49. Minneapolis: University of Minnesota Press.

Cannamela, Danila. 2019. *The Quiet Avant-Garde: Crepuscular Poetry and the Twilight of Modern Humanism*. Toronto: University of Toronto Press.

Carducci, Giosuè. 1942. "In riva al mare." In *Rime nuove*, 3. Turin: Einaudi.

Casid, Jill. 2005. *Sowing Empire: Landscape and Colonization*. Minneapolis: University of Minnesota Press.

Cassano, Franco. 2012. *Southern Thought and Other Essays on the Mediterranean*. New York: Fordham University Press.

Cassola, Carlo, and Luciano Bianciardi. 1956. *I minatori della Maremma*. Bari: Laterza.

Cattaneo, Giorgio. 2010. "*Meccanica celeste*: Maggiani, la bellezza salverà il mondo." Libreidee.org, September 5, 2010. http://www.libreidee.org/2010/05/meccanica-celeste-maggiani-la-bellezza-salvera-il-mondo.

Ceccagnoli, Patrizio. 2009. "'Necrofilia' e prosopopea della materia: La personificazione in Marinetti." *Annali d'Italianistica* 27:309–31.

———. 2013. "Marinetti e Venezia: Dal Romanticismo al feticismo." In *Venezianella e Studentaccio*, by F. T. Marinetti, edited by Paolo Valesio and Patrizio Ceccagnoli, 139–59. Milan: Mondadori.

Ceccardi, Roccatagliata Ceccardo. 1894. *Dai paesi dell'anarchia*. Deferrari

.it, November 2012. http://www.deferrari.it/FogliVia.htm.

———. (1905) 1982. "Apua mater." In *Tutte le poesie*, edited by Bruno Cicchetti and Eligio Imarisio, 252–53. Genoa: Sagep.

Cederna, Antonio. 1991. *Brandelli d'Italia*. Rome: Newton Compton.

Centemeri, Laura. 2010. "The Seveso Disaster Legacy." In *Nature and History in Modern Italy*, edited by Marco Armiero and Marcus Hall, 195–211. Athens: Ohio University Press.

Cesaretti, Enrico. 2013. *Fictions of Appetite: Alimentary Discourses in Modernist Italian Literature*. Oxford: Peter Lang.

Cigliana, Simona. 2002. *Futurismo esoterico*. Naples: Liguori.

City of Massa Carrara. n.d. "La capitale del marmo." Accessed April 2020. http://web.comune.carrara.ms.it/pagina1866_la-capitale-del-marmo.html.

Coda, Elena. 2002. "The Representation of the Metropolis in Scipio Slataper's *Il mio Carso*." *MLN* 117 (1): 153–73.

Cohen, Jeffrey. 2010. "Stories of Stone." *Postmedieval: A Journal of Medieval Cultural Studies* 1 (1/2): 56–63.

———, ed. 2014. *Inhuman Nature*. Washington, DC: Oliphaunt Books.

———. 2015. *Stone: An Ecology of the Inhuman*. Minneapolis: University of Minnesota Press.

Cohen, Jeffrey, and Lowell Duckert, eds. 2015. *Elemental Ecocriticism: Thinking with Earth, Air, Water, and Fire*. Minneapolis: University of Minnesota Press.

Consolo, Vincenzo. 1994. *L'olivo e l'olivastro*. Milan: Mondadori.

———. 1999. *Di qua dal faro*. Milan: Mondadori.

Coole, Diana, and Samantha Frost. 2010. *New Materialisms: Ontology, Agency, and Politics*. Durham: Duke University Press.

Costa degli Etruschi. n.d. "Costa degli Etruschi—Tuscany." Visit Costa degli Etruschi. Accessed April 2020. https://visitcostadeglietruschi.com/it/.

Cresswell, Tim. 2006. *On the Move: Mobility in the Modern Western World*. New York: Routledge.

Curi, Fausto. 2009. "Marinetti, il soggetto, la materia." *Annali d'Italianistica* 27:294–307.

Curino, Laura. 2012. "Malapolvere: Veleni e antidoti per l'invisibile." Raiplayradio.it. Accessed April 2020. https://www.raiplayradio.it/audio/2017/10/Raccontare-in-scena-4-Malapolvere-8beff44e-b800-4fa8-89ad-9a1eded7afe9.html.

Daly, Selena. 2013. "'The Futurist Mountains': Filippo Tommaso Marinetti's Experiences of Mountain Combat in the First World War." *Modern Italy* 18 (4): 323–38.

Danna, Daniela. 2017. "Expo 2015 in Milan: The Power of the Machine." *Journal of Political Ecology* 24:910–20.

Davì, Luigi. 1959. "I centauri." *Il Gatto Selvatico* 2 (5): 24–27.

De Boever, Arne. 2013. *Narrative Care: Biopolitics and the Novel*. New York: Bloomsbury.

DeLanda, Manuel. 1997. *A Thousand Years of Nonlinear History*. New York: Swerve Editions.

Deleuze, Gilles, and Félix Guattari. 1986. *Kafka: Toward a Minor Literature*. Minneapolis: University of Minnesota Press.

DeLoughrey, Elizabeth, and George B. Handley. 2011. *Postcolonial Ecologies: Literatures of the Environment*. Oxford: Oxford University Press.

Delville, Michel. 2008. *Food, Poetry, and the Aesthetic of Consumption: Eating the Avant-Garde*. New York: Routledge.

De Martino, Ernesto. 2002. *La fine del mondo: Contributo all'analisi delle apocalissi culturali*. Turin: Einaudi.

Dessì, Giuseppe. 1955. "Il fidanzato." *Il Gatto Selvatico* 2 (1): 15–16.

Di Martino, Loreadana, and Pasquale Verdicchio, eds. 2017. *Encounters with the Real in Contemporary Italian Literature and Cinema*. Newcastle upon Tyne: Cambridge Scholars.

Di Stefano, Paolo. 2011. *Viaggio in Italia: Un ritratto del paese nei racconti de "Il Gatto Selvatico" 1955-64*. Milan: BUR Rizzoli.

Dolphijn, Rick, and Iris van der Tuin. 2012. *New Materialism: Interviews and Cartographies*. Ann Arbor: Open Humanities Press.

Donati, Riccardo. 2016. "La grande astrazione: Su *Violazione* di Alessandra Sarchi." In *Ecosistemi letterari: Luoghi e paesaggi nella finzione novecentesca*, edited by Nicola Turi, 163–76. Florence: Florence University Press.

Donnarumma, Raffaele. 2012. "Tracciato del modernismo italiano." In *Sul modernismo italiano*, edited by Romano Luperini and Massimiliano Tortora, 13–38. Naples: Liguori.

Dürbeck, Gabriele, Caroline Schaumann, and Heather Sullivan. 2015. "Human and Non-human Agencies in the Anthropocene." *Ecozon@* 6 (1): 118–36.

Ente Nazionale Idrocarburi (ENI). n.d. *ENI e il Boom*. Archiviostorico.eni. Accessed October 2018. https://archiviostorico.eni.com/aseni/it/page/archive.

———. n.d. "Una raffineria in Lomellina." Archivio storico.eni. Accessed October 2018. https://archiviostorico.eni.com/aseni/it/page/archive.

Environmental Justice Atlas (EJAtlas). 2015. "Expo 2015, Italy." Ejatlas.org. Accessed April 2020. https://ejatlas.org/conflict/expo-2015-italy.

Esposito, Roberto. 2010. *Pensiero vivente: Origine e attualità della filosofia italiana*. Turin: Einaudi.

European Pollutant Release and Transfer Register (E-PRTR). 2017. "Pollutant Releases for Sarlux Srl/Saras (S.p.a)." prtr.eea.europa.eu. Accessed April 2020. https://prtr.eea.europa.eu/#/pollutantreleases.

Expo Milano. 2015. "Official Report, 2015." Expo2015.org. Accessed April 2020. http://www.expo2015.org/2018/06/07/expo-milano-2015-pubblica-il-suo-report-ufficiale/.

Falci, Eric. 2015. *The Cambridge Introduction to British Poetry, 1945–2010*. Cambridge: Cambridge University Press.

Falco, Giorgio. 2009. *L'ubicazione del bene*. Turin: Einaudi.

Farina, Almo. 2010. *Ecology, Cognition, and Landscape: Linking Natural and Social Systems*. Dordrecht: Springer.

Farr, Arnold. 2019. "Herbert Marcuse." Plato.stanford.edu. Accessed April 2020. https://plato.stanford.edu/archives/sum2019/entries/marcuse/.

Ferroni, Giulio. 2010. *Scritture a perdere: La letteratura negli anni zero*. Bari: Laterza.

Foundation for Environmental Education (FEE). 2019. "Bandiere blu dell'anno 2019." Bandiera Blu. Accessed April 2020. http://www.bandierablu.org/common/blueflag.asp?anno=2019&tipo=bb.

Gadda, Carlo Emilio. 1960. "Il pozzo n. 14." *Il Gatto Selvatico* 1 (6): 30–32.

———. 1964. *Le meraviglie d'Italia*. Turin: Einaudi.

Gersdorf, Catrin, and Sylvia Mayer. 2006. "Nature in Literary and Cultural Studies: Defining the Subject of Ecocriticism—an Introduction." In *Nature in Literary and Cultural Studies: Transatlantic Conversations on Ecocriticism*, edited by Catrin Gersdorf and Sylvia Mayer, 9–21. Amsterdam: Rodopi.

Ghosh, Amitav. (1992) 2005. "Petrofiction: The Oil Encounter and the Novel." In *Incendiary Circumstances: A Chronicle of the Turmoil of Our Times*, 138–51. Boston: Houghton Mifflin.

———. 2017. "From Bombay to Canton: Traveling the Indian Ocean Opium Route." Lecture given at the University of Virginia, April 26, 2017.

Giancotti, Matteo. 2013. *Andrea Zanzotto: Luoghi e paesaggi*. Milan: Bompiani.

Giannini, Sirio. 1958. *La valle bianca*. Milan: Mondadori.

———. 1971. *Dove nasce il fiume*. Bologna: Massimiliano Bono Editore.

Glotfelty, Cheryll. 1996. "Introduction: Literary Studies in an Age of Environmental Crisis." In *The Ecocriticism Reader: Landmarks in Literary Ecology*, edited by Cheryill Glotfelty and Harold Fromm, xv–xxxvii. Athens: University of Georgia Press.

Goretti, Maria. 1941. *La donna e il futurismo*. Verona: La Scaligera.

———. (1941) 2009. "Song of Petroleum." In *Futurism: An Anthology*, edited by Lawrence Rainey and Laura Wittman, 476–79. New Haven: Yale University Press.

Gregson, Nicky, Helen Watkins, and Melania Calestani. 2010. "Inextinguishable Fibers: Demolition and the Vital Materialisms of Asbestos."

Guarnieri, Silvio. 1979. Introduction to *Moscardino; Il servitore del diavolo; Il volto santo*, by Enrico Pea, v–xxii. Turin: Einaudi.

Guattari, Félix. 1989. "The Three Ecologies." *New Formations* 8:131–47.

Guattari, Félix, and Gilles Deleuze. 1983. "What Is a Minor Literature?" *Mississippi Review* 11 (3): 13–33.

Halliwell, Martin, and Andy Mousley. 2003. *Critical Humanisms: Humanist/Anti-humanist Dialogues*. Edinburgh: Edinburgh University Press.

Hamilton, Clive. 2015. "The Theodicy of the 'Good Anthropocene.'" *Environmental Humanities* 7:233–38.

Harkness, Rachel, Cristián Simonetti, and Judith Winter. 2018. "Concretes Speak: A Play in One Act." In *Future Remains: A Cabinet of Curiosities for the Anthropocene*, edited by Gregg Mitman, Marco Armiero, and Robert S. Emmett, 29–39. Chicago: University of Chicago Press.

Härmänmaa, Marja. 2009. "Futurism and Nature: The Death of the Great Pan." In *Futurism and the Technological Imagination*, edited by Günter Berghaus, 337–60. Amsterdam: Rodopi.

Harvey, David. 2000. *Spaces of Hope*. Berkeley: University of California Press.

Heise, Ursula. 2008. *Sense of Place and Sense of Planet: The Environmental Imagination of the Global*. Oxford: Oxford University Press.

———. 2016. "The Environmental Humanities and the Futures of the Human." *New German Critique* 43:21–31.

Holm, Poul, Joni Adamson, Hsinya Huang, Lars Kirdan, Sally Kitch, Iain McCalman, James Ogude et al. 2015. "Humanities for the Environment: A Manifesto for Research and Action." *Humanities* 4 (4): 977–92. https://doi.org/10.3390/h4040977.

Houser, Heather. 2014. *Ecosickness in Contemporary U.S. Fiction*. New York: Columbia University Press.

Iovino, Serenella. 2009. "Naples 2008, or The Waste Land: Trash, Citizenship, and an Ethic of Narration." *Neohelicon* 36 (2): 335–46.

———. 2012a. "Material Ecocriticism: Matter, Text, and Posthuman Ethics." In *Literature, Ecology, Ethics*, edited by Timo Müller and Michael Sauter, 51–68. Heidelberg: Winter Verlag.

———. 2012b. "Restoring the Imagination of Place: Narrative Reinhabitation and the Po Valley." In *The Bioregional Imagination: Literature, Ecology, and Place*, edited by Tom Lynch, Cheryll Glotfelty, and Karla Ambruster, 100–117. Athens: University of Georgia Press.

———. 2012c. "Steps to a Material Ecocriticism: The Recent

Literature About the New Materialisms and Its Implications for Ecocritical Theory." *Ecozon@* 3 (2): 134–45.

———. 2014. "Bodies of Naples: Stories, Matter, and the Landscapes of Porosity." In *Material Ecocriticism*, edited by Serenella Iovino and Serpil Oppermann, 971–113. Bloomington: Indiana University Press.

———. 2016a. *Ecocriticism and Italy: Ecology, Resistance, and Liberation*. London: Bloomsbury.

———. 2016b. "Posthumanism in Literature and Ecocriticism." *Relations: Beyond Anthropocentrism* 4 (1): 1120.

Iovino, Serenella, and Serpil Oppermann, eds. 2012. "Theorizing Material Ecocriticism: A Dyptich." *Interdisciplinary Studies in Literature and the Environment* 19 (3): 448–75.

———, eds. 2014. *Material Ecocriticism*. Bloomington: Indiana University Press.

Istituto Superiore per la Protezione e la Ricerca Ambientale (ISPRA). 2017. *Consumo di suolo, dinamiche territoriali e servizi ecosistemici*. Isprambiente.gov.it, June 2017. http://www.isprambiente.gov.it/en/publications/reports?b_start:int=40.

Italcementi. 2015. "Cemento biodinamico per Expo 2015." Italcementi.it. Accessed April 2020. https://www.italcementi.it/it/palazzo-italia-expo-2015.

Jackson, B. John. 1984. *Discovering the Vernacular Landscape*. New Haven: Yale University Press.

James, Erin. 2015. *The Storyworld Accord: Econarratology and Postcolonial Narratives*. Lincoln: University of Nebraska Press.

Jansen, Monica. 2017. "The Uses of Affective Realism in Asbestos Narratives: Prunetti's *Amianto* and Valenti's *La fabbrica del panico*." In *Encounters with the Real in Contemporary Italian Literature and Cinema*, edited by Loreadana Di Martino and Pasquale Verdicchio, 3–27. Newcastle upon Tyne: Cambridge Scholars.

Jansen, Monica, and Maria Bonaria Urban. 2017. "Raccontare la giustizia." *Forum Italicum* 51 (1): 1–11.

Kerényi, Karol. 2014. *Rapporto con il divino e altri saggi*. Milan: Bompiani.

Kern, Robert. 2000. "Ecocriticism: What Is It Good For?" *ISLE: Interdisciplinary Studies in Literature and Environment* 7 (1): 932.

Kolbert, Elizabeth. 2014. *The Sixth Extinction: An Unnatural History*. New York: Henry Holt.

Kroll, Gary. 2018. "Snarge." In *Future Remains: A Cabinet of Curiosities for the Anthropocene*, edited by Gregg Mitman, Marco Armiero, and Robert S. Emmett, 81–88. Chicago: University of Chicago Press.

Krumm, Ermanno. 2003. *Animali e uomini*. Turin: Einaudi.

Legambiente. 2012. *Cemento Spa: Mafie, corruzione e abusivismo edilizio; Numeri, storie e misfatti di chi sta saccheggiando il Nord*. Legambiente.it. Accessed April 2020. http://www.legambiente.it/sites/default/files/docs/dossier_cemento_spa_def_0.pdf.

Lehtimäki, Markku. 2013. "Natural Environments in Narrative Contexts: Cross-pollinating Ecocriticism and Narrative Theory." *Storyworlds: A Journal of Narrative Studies* 5:119–41.

LeMenager, Stephanie. 2012. "The Aesthetics of Petroleum, After Oil!" *American Literary History* 24 (1): 5986.

———. 2014. *Living Oil: Petroleum Culture in the American Century*. Oxford: Oxford University Press.

Leopardi, Giacomo. 1983. *The Moral Essays*. Translated by Patrick Creagh. New York: Columbia University Press.

Levi, Primo. 1984. *The Periodic Table*. Translated by Raymond Rosenthal. New York: Schocken Books. Originally published as *Il sistema periodico* (Turin: Einaudi, 1975).

———. 1989. *Other People's Trades*. Translated by Raymond Rosenthal. New York: Summit Books. Originally published as *L'altrui mestiere* (Turin: Einaudi, 1985).

Lindgren, Mia, and Gail Phillips. 2016. "Asbestos Memories: Journalistic 'Mediation' in Mediated Prospective Memory." In *Memory in a Mediated World: Remembrance and Reconstruction*, edited by Andrea Hajek, Christine Lohmeier, and Christian Pentzold, 158–78. New York: Palgrave Macmillan.

Lindström, Karl, Hannes Palang, and Kalevi Kull. 2013. "Semiotics of Landscape." In *The Routledge Companion to Landscape Studies*, edited by Peter Howard, Ian Thompson, Emma Waterton, and Mick Atha, 97–107. London: Routledge.

Locatelli, Luca, and Sam Anderson. 2017. "The Majestic Marble Quarries of Northern Italy." *New York Times*, July 26, 2017. https://www.nytimes.com/2017/07/26/magazine/the-majestic-marble-quarries-of-northern-italy.html.

Lopez, Barry. 1989. "The American Geographies." *Orion* (Autumn): 52–61.

Lucarelli, Carlo. 2012. "Amianto: Le morti silenziose." Blunotte.rai.it, February 13, 2012. https://blunotte.rai.it/dl/portali/site/puntata/ContentItem-99ba6c31-c0b0-4a0f-9cd4-e28a5c0a2933.html.

Luisetti, Federico, John Pickles, and Wilson Kaiser, eds. 2016. *The Anomie of the Earth: Philosophy, Politics, and Autonomy in Europe and the Americas*. Durham: Duke University Press.

Macdonald, Graeme. 2012. "Oil and World Literature." *American Book Review* 33 (3): 7–31.

Maggiani, Maurizio. 2010. *Meccanica celeste*. Milan: Feltrinelli.

Magnaghi, Alberto. 2005. *The Urban Village*. New York: Zed Books.

Malm, Andreas. 2017. "'This Is the Hell That I Have Heard Of': Some Dialectical Images in Fossil Fuel Fiction." *Forum for Modern Language Studies* 53 (2): 121–41.

Maltagliati, Silvia, and Antongiulio Barbaro. 2012. "Indagine sociale sulle maleodoranze intorno all'area Picchianti, Livorno." Arpat.toscana.it. Accessed April 2020. http://www.arpat.toscana.it/documentazione/report/indagine-sociale-sulle-maleodoranze-intorno-all2019area-picchianti-livorno-analisi-delle-schede-di-rilevazione.

Mangano, Daniel. 1996. "Paesaggio e futurismo." *Narrativa* 9 (February): 55–66.

Maniowska, Katarzyna. 2012. "La società scomposta: il progresso in Sardegna ne *Il parroco di Arasolè* di Francsco Masala." *Romanica Cracoviensia* 12, 314–27.

Mao, Douglas. 1998. *Solid Objects: Modernism and the Test of Production*. Princeton: Princeton University Press.

Maran, Timo. 2010. "An Ecosemiotic Approach to Nature Writing." *PAN: Philosophy, Activism, Nature* 7:79–87.

Marchesini, Roberto. 2002. *Posthuman: Verso nuovi modelli di esistenza*. Turin: Bollati-Boringhieri.

———. 2014. "Mimesis: The Heterospecific as Ontopoietic Epiphany." In *Thinking Italian Animals: Human and Posthuman in Modern Italian Literature and Film*, edited by Deborah Amberson and Elena Past, xiii–xxxvi. New York: Palgrave Macmillan.

———. 2016. *Alterità: L'identità come relazione*. Modena: Mucchi.

Marinetti, F. T. 1905. *Le Roi Bombance*. Paris: Sociéte Mercure de France.

———. 1912. *Le monoplan du pape*. Paris: Bibliothèque Internationale d'édition E. Sansot.

———. 1960a. *Il teatrino dell'amore*. In *Teatro*, vol. 2, edited by Giovanni Calendoli, 341–45. Rome: Vito Bianco Editore.

———. 1960b. *Ricostruire l'Italia con architettura futurista Sant'Elia*. In *Teatro*, vol. 3, edited by Giovanni Calendoli, 509–602. Rome: Vito Bianco Editore.

———. 1960c. *Vengono*. In *Teatro*, vol. 2, edited by Giovanni Calendoli, 283–85. Rome: Vito Bianco Editore.

———. 1960d. *Vulcano: 8 sintesi incatenate*. In *Teatro*, vol. 2, edited by Giovanni Calendoli, 137–206. Rome: Vito Bianco Editore.

———. 1969a. *La grande Milano tradizionale e futurista*. In *La grande Milano tradizionale e futurista. Una sensibilità italiana nata in Egitto*, edited by Luciano De Maria, 3–198. Milan: Mondadori.

———. 1969b. *Una sensibilità italiana nata in Egitto*. In *La grande Milano tradizionale e futurista. Una sensibilità italiana nata in Egitto*, edited by Luciano De Maria, 199–334. Milan: Mondadori.

———. 1990a. "Distruzione della sintassi. Immaginazione senza fili. Parole in libertà." In *Teoria e invenzione futurista*, edited by Luciano De Maria, 65–80. Milan: Mondadori.

———. 1990b. "Fondazione e manifesto del futurismo." In *Teoria e invenzione futurista*, edited by Luciano De Maria, 7–14. Milan: Mondadori.

———. 1990c. *Il fascino dell'Egitto*. In *Teoria e invenzione futurista*, edited by Luciano De Maria, 1051–91. Milan: Mondadori.

———. 1990d. "Il poema non umano dei tecnicismi." In *Teoria e invenzione futurista*, edited by Luciano De Maria, 1141–94. Milan: Mondadori.

———. 1990e. "Il Tattilismo." In *Teoria e invenzione futurista*, edited by Luciano De Maria, 159–66. Milan: Mondadori.

———. 1990f. "L'aeropoema del Golfo della Spezia." In *Teoria e invenzione futurista*, edited by Luciano De Maria, 1095–137. Milan: Mondadori.

———. 1990g. "La guerra, sola igiene del mondo." In *Teoria e invenzione futurista*, edited by Luciano De Maria, 290–341. Milan: Mondadori.

———. 1990h. "La nuova religione-morale della velocità." In *Teoria e invenzione futurista*, edited by Luciano De Maria, 130–38. Milan: Mondadori.

———. 1990i. "Lo splendore geometrico e meccanico e la sensibilità numerica." In *Teoria e invenzione futurista*, edited by Luciano De Maria, 98–107. Milan: Mondadori.

———. 1990j. "Manifesto tecnico della letteratura futurista." In *Teoria e invenzione futurista*, edited by Luciano De Maria, 46–54. Milan: Mondadori.

———. 1990k. *Spagna veloce e toro futurista*. In *Teoria e invenzione futurista*, edited by Luciano De Maria, 1013–50. Milan: Mondadori.

———. 1990l. "Uccidiamo il chiaro di luna." In *Teoria e invenzione futurista*, edited by Luciano De Maria, 14–26. Milan: Mondadori.

———. 1998a. *Mafarka the Futurist: An African Novel*. Translated by Carol Diethe and Steve Cox. London: Middlesex University Press.

———. 1998b. *The Untameables*. Translated by Jeremy Parzen. Los Angeles: Sun and Moon Press.

———. 2003a. "11 baci a Rosa di Belgrado." In *Novelle colle labbra tinte*, edited by Domenico Cammarota, 28–58. Florence: Vallecchi.

———. 2003b. "La logica di Ahmed Bey." In *Novelle colle labbra tinte*, edited by Domenico Cammarota, 103–9. Florence: Vallecchi.

———. 2003c. "Meandri di un rio nella foresta brasiliana." In *Novelle colle labbra tinte*, edited by Domenico Cammarota, 69–76. Florence: Vallecchi.

———. 2003d. *Novelle colle labbra tinte*. Edited by Domenico Cammarota. Florence: Vallecchi.

———. 2013. *Venezianella e Studentaccio*. Milan: Mondadori.

Marinetti, F. T., and Luigi Colombo. 1989. *The Futurist Cookbook*. Translated by Suzanne Brill. London: Trefoil. Originally published as *La cucina futurista* (Milan: Longanesi, 1986).

Masala, Francesco. 2001. *Il parroco di Arasolè*. Nuoro: Il Maestrale.

Mazza Galanti, Carlo. 2012. "Riflessioni su *Il contro in testa*." *Minima & Moralia* (blog), August 25, 2012. http://www.minimaetmoralia.it/wp/riflessioni-su-il-contro-in-testa.

Mazzotta, Massimiliano. 2009. *Oil: La forza devastante del petrolio; La dignità del popolo Sardo*. Vimeo. Accessed April 2020. https://vimeo.com/17046133.

Menduni, Enrico. 1999. *L'autostrada del Sole*. Bologna: Il Mulino.

Meschiari, Matteo. 2008. *Sistemi selvaggi: Antropologia del paesaggio scritto*. Palermo: Sellerio.

Mezzadra, Sandro, and Brett Neilson. 2017. "On the Multiple Frontiers of Extraction: Excavating Contemporary Capitalism." *Cultural Studies* 31 (2/3): 185–204. http://dx.doi.org/10.1080/09502386.2017.1303425.

Mignolo, Walter. 2016. "Sustainable Development or Sustainable Economies? Ideas Towards Living in Harmony and Plenitude." Doc Research Institute. Accessed April 2020. https://doc-research.org/wp-content/uploads/2016/10/Mignolo-Final-Text-Rhodes-Forum.pdf.

———. 2015. *La terra bianca: Marmo, chimica e altri disastri*. Rome: Laterza.

Minervino, Mauro F. 2010. *Statale 18*. Rome: Fandango.

———. 2011. "*Statale 18* di Mauro Francesco Minervino." YouTube, October 17, 2011. https://www.youtube.com/watch?v=pbKAsvRlVic.

Mistiaen, Veronique, and Chiara Briganti. 2015. "Michelangelo's Marble Is Being Sold Cheap by Industrialists." *Newsweek*, March 27, 2015. http://www.newsweek.com/2015/04/03/bin-ladens-and-tuscan-city-destroyed-marble-317224.html.

Mitman, Gregg, Marco Armiero, and Robert S. Emmett, eds. 2018. *Future Remains: A Cabinet of Curiosities for the Anthropocene*. Chicago: University of Chicago Press.

Moroni, Mario. 2011. "Beyond Matter: Futurism and Occultist Practices." In *Futurismo: Impact and Legacy*, edited by Giuseppe Gazzola, 117–24. Stony Brook, NY: Forum Italicum.

Morton, Timothy. 2010. *The Ecological Thought*. Cambridge: Harvard University Press.

Mossano, Silvana. 2010. *Malapolvere: Una città si ribella ai "signori" dell'amianto*. Alba: Edizioni Sonda.

Mossano, Silvana, and Michele Brambilla. 2012. *Morire d'amianto: Il caso Eternit; La fabbrica, le vittime, la giustizia*. Turin: La Stampa.

Mussolini, Benito. 2012. "Discorso di Reggio Emilia, 30 Ottobre 1926." Bibliotecafascista, March 3, 2012. http://bibliotecafascista.blogspot.com/2012/03/discorso-di-reggio-emilia-30-ottobre.html.

Neimanis, Astrida, Cecilia Åsberg, and Johan Hedrén. 2015. "Four Problems, Four Directions for Environmental Humanities: Toward Critical Posthumanities for the Anthropocene." *Ethics and the Environment* 20 (1): 67–97.

Newman, Richard. 2012. "Darker Shades of Green: Love Canal, Toxic Autobiography, and American Environmental Writing." In *Histories of the Dustheap: Waste, Material Cultures, Social Justice*, edited by Stephanie Foote and Elizabeth Mazzolini, 21–47. Cambridge: MIT Press.

Nixon, Rob. 2011. *Slow Violence and the Environmentalism of the Poor*. Cambridge: Harvard University Press.

Nordhaus, Ted, and Michael Shellenberger. 2015. "An Ecomodernist Manifesto: From the Death of Environmentalism to the Birth of Ecomodernism." Breakthrough Institute, April 15, 2015. https://thebreakthrough.org/.

Ogden, Laura. 2011. *Swamplife: People, Gators, and Mangroves Entangled in the Everglades*. Minneapolis: University of Minnesota Press.

Oliveira, Henrique. 2016. *Transcorredor*. Henriqueoliveira.com. Accessed April 2020. http://www.henriqueoliveira.com/portu/comercio_i.asp?flg_Lingua=1&cod_menu_obras=1&cod_Serie=8&cod_Artista=1.

Oppermann, Serpil. 2017. "Nature's Narrative Agencies as Compound Individuals." *Neohelicon* 44 (2): 283–95. https://link.springer.com/article/10.1007/s11059-017-0394-9.

———. 2018. "Ecomaterialism." In *Posthuman Glossary*, edited by Rosi Braidotti and Maria Hlavajova, 120–23. London: Bloomsbury.

Owen, David. 2017. "The End of Sand." *New Yorker*, May 29, 2017, 28–33.

Palazzeschi, Aldo. 1910. *L'incendiario*. Milan: Edizioni Futuriste di "Poesia."

Palumbo, Patrizia, ed. 2003. *A Place in the Sun: Africa in Italian Colonial Culture from Post-unification to the Present*. Berkeley: University of California Press.

Pasolini, Pier Paolo. 1965. "Love Meetings (1964)." IMDb. Accessed April 2020. https://www.imdb.com/title/tt0057960/.

———. (1972) 1992. *Petrolio*. Turin: Einaudi.

———. 1975. "Il romanzo delle stragi." In *Scritti corsari*, 111–17. Milan: Garzanti.

Pasolini, Pier Paolo, and Giovanni Bonfanti. 1972. *12 Dicembre*.

YouTube. Accessed April 2020. https://www.youtube.com/watch?v=K2Rg_5YZV08.
Past, Elena. 2016. "Mediterranean Ecocriticism: The Sea in the Middle." In *Handbook of Ecocriticism and Cultural Ecology*, edited by Hubert Zapf, 368–84. Berlin: De Gruyter.
Past, Elena, and Deborah Amberson. 2016. "Gadda's Pasticciaccio and the Knotted Posthuman Household." *Relations* 4 (1): 65–79.
Pea, Enrico. 1910. *Le fole*. Pescara: Industrie grafiche.
———. 1979. *Moscardino; Il servitore del diavolo; Il volto santo*. Turin: Einaudi.
———. 1982. *Vita in Egitto*. Milan: Mondadori.
Perniola, Mario. 2003. "On Italian Neonaturalism." Agalmarivista. Accessed April 2020. https://www.agalmarivista.org/fascicoli/agalma-4-natura-coltura-cultura/.
———. 2004. *The Sex Appeal of the Inorganic: Philosophies of Desire in the Modern World*. New York: Continuum.
Pertici, Roberto. 2010. "Ceccardo Roccatagliata Ceccardi fra mito e realtà storica." In *La Repubblica di Apua*, edited by Stefano Bucciarelli, Marcello Ciccuto, and Antonella Serafini, 27–33. Florence: Maschietto Editore.
Petitto, Mario. 2012. *L'età del cemento*. Vimeo. Accessed April 2020. https://vimeo.com/178545681.
Pickering, Andrew. 1995. *The Mangle of Practice: Time, Agency, and Science*. Chicago: University of Chicago Press.
Pickering-Iazzi, Robin. 1995. *Mothers of Invention: Women, Italian Fascism, and Culture*. Minneapolis: University of Minnesota Press.
Pinkus, Karen. 2012. "Selling Gasoline in Autarchic Italy." In *Figura umana: Normkonzepte der Menschendarstellung in der italienischen Kunst 1919-1939*, 151–60. Petersberg: Michael Imhof Verlag.
———. 2014. "Pasolini's Petrolio: Fossil Fuel, Chaotic Desire, Anthropocene Narratives." Program in Critical Theory, September 18, 2014. http://criticaltheory.berkeley.edu/2014-events.
———. 2016. *Fuel: A Speculative Dictionary*. Minneapolis: University of Minnesota Press.
Piovene, Guido. 1957. *Viaggio in Italia*. Milan: Mondadori.
Pirandello, Luigi. 1992. *One, No One, and One Hundred Thousand*. Translated by William Weaver. New York: Marsilio. Originally published as *Uno, nessuno e centomila* (Milan: Garzanti, 1983).
———. 1994a. "Ciàula scopre la luna." In *Novelle per un anno*, vol. 2, 1278–85. Florence: Giunti.
———. 1994b. "Formalità." In *Novelle per un anno*, vol. 1, 116–37. Florence: Giunti.
———. 1994c. "Fuoco alla paglia." In *Novelle per un anno*, vol. 1, 278–87. Florence: Giunti.

———. 1994d. "Il fumo." In *Novelle per un anno*, vol. 1, 45–78. Florence: Giunti.
Plumwood, Val. 1993. *Feminism and the Mastery of Nature*. New York: Routledge.
Poggi, Christine. 1997. "Dreams of Metallized Flesh: Futurism and the Masculine Body." *Modernism/Modernity* 4 (3): 1943.
———. 2009. *Inventing Futurism: The Art and Politics of Artificial Optimism*. Princeton: Princeton University Press.
Ponzanesi, Sandra. 2012. "The Postcolonial Turn in Italian Studies: European Perspectives." In *Postcolonial Italy: Challenging National Homogeneity*, edited by Cristina Lombardi-Diop and Caterina Romeo, 51–69. New York: Palgrave Macmillan.
Posman, Sarah, Anne Reverseau, and David Ayers, eds. 2013. *The Aesthetics of Matter*. Berlin: De Gruyter.
Pound, Ezra. 1941. "From His Radio Speeches." Archipelago Books. Accessed April 2020. https://archipelagobooks.org/book/moscardino/.
Princeton University Press. n.d. Book overview of *Solid Objects: Modernism and the Test of Production*, by Douglas Mao. Accessed April 2020. https://press.princeton.edu/books/ebook/9781400822706/solid-objects.
Prunetti, Alberto. 2012. *Amianto: Una storia operaia*. Rome: Edizioni Alegre.

Pugno, Laura. 2012. "Storie di personaggi contaminati dalla corruzione o dall'acquiescenza." *Il Manifesto*, June 27, 2012, 11.
Quasimodo, Salvatore. 1966. "A un poeta nemico." In *Il falso e vero verde*, 45. Milan: Mondadori.
Raine, Anne. 2014. "Ecocriticism and Modernism." In *The Oxford Handbook of Ecocriticism*, edited by Greg Garrard, 98–117. Oxford: Oxford University Press.
Rainey, Lawrence, Christine Poggi, and Laura Wittman, eds. 2009. *Futurism: An Anthology*. New Haven: Yale University Press.
Re, Lucia. 1989. "Futurism and Feminism." *Annali d'Italianistica* 7:253–72.
———. 2003. "Alexandria Revisited: Colonialism and the Egyptian Works of Enrico Pea and Giuseppe Ungaretti." In *A Place in the Sun: Africa in Italian Colonial Culture from Post-unification to the Present*, edited by Patrizia Palumbo, 163–96. Berkeley: University of California Press.
Rigby, Kate. 2014. "Confronting Catastrophe: Ecocriticism in a Warming World." In *The Cambridge Companion to Literature and the Environment*, edited by Louise Westling, 212–25. Cambridge: Cambridge University Press.
Rigell, Laura. 2013. "Sicily's Socialist Fasci Unite for Workers' Rights, Italy, 1893–1894." Global Nonviolent Action Database, February 3, 2013. https://nvdatabase.swarthmore.edu/content/sicily

-socialist-fasci-unite-workers-rights-italy-1893-1894.

Rovelli, Marco. 2014. *Il contro in testa: Gente di marmo e d'anarchia*. Rome: Laterza.

Rust, Stephen, Salma Monani, and Sean Cubitt, eds. 2013. *Ecocinema: Theory and Practice*. New York: Routledge.

Ryan, Derek. 2015. "Following Snakes and Moths: Modernist Ethics and Posthumanism." *Twentieth-Century Literature* 61 (3): 287–304.

Sant'Elia, Antonio. 1914. "Manifesto dell'architettura futurista." Tiscali.it. Accessed April 2020. http://web.tiscali.it/antonio_santelia/manifesto.htm.

Sarchi, Alessandra. 2012. *Violazione*. Turin: Einaudi.

———. 2014. *L'amore normale*. Turin: Einaudi.

———. 2017. *La notte ha la mia voce*. Turin: Einaudi.

Saviano, Roberto. 2015. "Così il Nobel della realtà rivoluziona la letteratura." *La Repubblica*, October 12, 2015. https://www.repubblica.it/cultura/2015/10/12/news/nobel_aleksievic-124875631/.

Schliephake, Christopher. 2015. *Urban Ecologies: City Space, Material Agency, and Environmental Politics in Contemporary Culture*. Lanham, MA: Lexington Books.

Schnapp, Jeffrey. 2004. "Rayon/Marinetti." In *Science and Literature in Italian Culture from Dante to Calvino: A Festschrift for Patrick Boyde*, edited by Pierpaolo Antonello and Simon A. Gilson, 226–53. Oxford: European Humanities Research Center.

Schoonover, Karl. 2013. "Documentaries Without Documents? Ecocinema and the Toxic." *Necsus* 2 (2): 483–507.

Schuster, Joshua. 2015. *The Ecology of Modernism: American Environments and Avant-Garde Poetics*. Tuscaloosa: University of Alabama Press.

Sciascia, Leonardo. 1964. "Gela: Realtà e condizione umana." *Il Gatto Selvatico* 3 (10): 17–20.

———. 1985. "The Wine-Dark Sea." In *The Wine-Dark Sea*. Translated by Avril Bardoni. Oxford: Carcanet Press. Originally published as "Il mare colore del vino" (Turin: Einaudi, 1973).

Sciascia, Leonardo, and Giuseppe Ferrara. 1964. *Gela antica e nuova* a.k.a. *I miti e il petrolio*. YouTube. Accessed April 2020. https://www.youtube.com/watch?v=UmvsnUAg7r4.

Seger, Monica. 2015. *Landscapes in Between: Environmental Change in Modern Italian Literature and Film*. Toronto: University of Toronto Press.

———. 2017. "Narrating Dioxin: Laura Conti's *A Hare with the Face of a Child*." *ISLE* 24 (1): 47–65.

———. 2018. "Thinking Through Taranto: Toxic Embodiment, Eco-catastrophe, and the Power of Narrative." In *Italy and the Environmental Humanities: Landscapes, Natures, Ecologies*, edited by Serenella Iovino, Enrico Cesaretti, and Elena Past, 184–93. Charlottesville: University of Virginia Press.

Settis, Salvatore. 2010. *Paesaggio, costituzione, cemento: La battaglia per l'ambiente contro il degrado civile*. Turin: Einaudi.

Sheller, Mimi. 2014. *Aluminum Dreams: The Making of Light Modernity*. Cambridge: MIT Press.

Slack, Jennifer Daryl, and J. Macgregor Wise. 2005. *Culture and Technology: A Primer*. New York: Peter Lang.

Slataper, Scipio. (1912) 1988. *Il mio Carso*. Trieste: Edizioni Italo Svevo.

Snyder, Gary. 1995. *A Place in Space*. Berkeley, CA: Counterpoint.

Solnit, Rebecca. 2009. *A Paradise Built in Hell: The Extraordinary Communities That Arise in Disaster*. New York: Penguin.

Somigli, Luca. 2016. "Nature and the Madman: Mysticism and Modernism in Pirandello's Fiction." In *Modernism and the Avant-Garde Body in Spain and Italy*, edited by Maria Truglio and Nicolás Fernández-Medina, 132–50. New York: Routledge.

Squarcia, Francesco. 1959. "Un pedone al Motèl." *Il Gatto Selvatico* 1 (5): 15–17.

Stella, Gian Antonio. 2015. "Da risorsa a minaccia al paesaggio: La Toscana alla disfida del marmo." *Il Corriere della Sera*, March 17, 2015.

Sudan, Rajani. 2016. *The Alchemy of Empire: Abject Materials and the Technologies of Colonialism*. New York: Fordham University Press.

Sullivan, Heather. 2014. "Dirty Traffic and the Dark Pastoral in the Anthropocene: Narrating Refugees, Deforestation, Radiation, and Melting Ice." *Literatur für Leser* 73 (2): 83–97.

———. 2015. "The Dark Pastoral." EASLCE Webinar, May 21, 2015. http://www.easlce.eu/webinar-transcript-the-dark-pastoral-with-prof-heather-i-sullivan.

———. 2017. "Material Ecocriticism and the Petro-text." In *The Routledge Companion to the Environmental Humanities*, edited by Ursula K. Heise, Jon Christensen, and Michelle Nieman, 414–23. London: Routledge.

Szeman, Imre. 2010. "The Cultural Politics of Oil: On Lessons of Darkness and Black Sea Flies." *Polygraph* 22:33–45.

———. 2011. "Literature and Energy Futures." *PMLA* 126 (2): 323–25.

———. 2012. "Introduction to Focus: Petrofictions." *American Book Review* 33 (3): 3–3.

———. 2014. "Crude Aesthetics: The Politics of Oil Documentaries." In *Oil Culture*, edited by Ross Barrett and Daniel Worden, 350–65. Minneapolis: University of Minnesota Press.

Terrosi, Roberto. 2009. "Futurismo e postumano." *Annali d'Italianistica* 27:263–73.

Thompson, Kara. 2019. *Blanket*. New York: Bloomsbury Academic.

Tomasi, Franco. 2014. "Immagini della megalopolis padana in *L'ubicazione del bene* di Giorgio Falco." In *La geografia del racconto: Sguardi interdisciplinari sul paesaggio urbano nella narrativa italiana contemporanea*, edited by Davide Papotti and Franco Tomasi, 91–111. Brussels: Peter Lang.

Trevisan, Vitaliano. 2010. *Tristissimi giardini*. Bari: Laterza.

Turri, Eugenio. 1979. *Semiologia del paesaggio italiano*. Milan: Longanesi.

———. 1998. *Il paesaggio come teatro: Dal territorio vissuto al territorio rappresentato*. Venice: Marsilio.

Turri, Eugenio, and Mimmo Jodice. 2001. *Gli iconemi: Storia e memoria del paesaggio*. Milan: Electa.

Tutter, Adele. 2015. "Giuseppe Penone: Indistinti confini / Indistinct Boundaries." Brooklyn Rail. Accessed April 2020. http://brooklynrail.org/2015/04/artseen/giuseppe-penone-indistinti-confini-indistinct-boundaries.

Valdemarca, Gioia. 2014. "Le città invivibili. Visioni di zone industriali." In *La geografia del racconto: Sguardi interdisciplinari sul paesaggio urbano nella narrativa italiana contemporanea*, edited by Davide Papotti and Franco Tomasi, 81–89. Brussels: Peter Lang.

Valenti, Stefano. 2013. *La fabbrica del panico*. Milan: Feltrinelli.

Valesio, Paolo. 2004. "Foreword: *After the Conquest of the Stars*." In *Italian Modernism: Italian Culture Between Decadentism and Avant-Garde*, edited by Luca Somigli and Mario Moroni, i–xxxiii. Toronto: University of Toronto Press.

———. 2013. "Il portasigarette ritrovato." In *Venezianella e Studentaccio*, by F. T. Marinetti, vii–liv. Milan: Mondadori.

Vallerani, Francesco, and Mauro Varotto. 2005. *Il grigio oltre le siepi: Geografie smarrite e racconti del disagio in Veneto*. Portogruaro: Nuova Dimensione.

Varotto, Mauro. 2014. "Geografie dell'abbandono nella periferia diffusa: *I quindicimila passi* di Vitaliano Trevisan." In *La geografia del racconto: Sguardi interdisciplinari sul paesaggio urbano nella narrativa italiana contemporanea*, edited by Davide Papotti and Franco Tomasi, 113–30. Brussels: Peter Lang.

Verdicchio, Pasquale, ed. 2016. *Ecocritical Approaches to Italian Culture and Literature: The Denatured Wild*. Lanham, MA: Lexington Books.

Vinci, Simona. 2007. *Rovina*. Milano: Edizioni Ambiente.

Viola, E. Gianni. 2010. "'L'ossessione lirica della materia." Academia.edu, February 2010. https://www.academia.edu/6016676/Lossessione_lirica_della_materia.

von Bismark, Julius, and Julian Charrière. 2014. *Clockwork*. julian-charriere.net. Accessed April 2020. http://julian-charriere.net/projects/clockwork.

Watts, Christopher, ed. 2013. *Relational Archaeologies: Humans, Animals, Things*. New York: Routledge.

Wenzel, Jennifer. 2014. "How to Read for Oil." *Resilience: A Journal of the Environmental Humanities* 1 (3): 156–61.

Withers, Jeremy. 2017. "Introduction: The Ecologies of Mobility." *ISLE: Interdisciplinary Studies in Literature and Environment* 24 (1): 66–74.

Wu Ming 1. 2008. "New Italian Epic 2.0." Carmilla, September 15, 2008. https://www.carmillaonline.com/2008/09/15/new-italian-epic-20.

Wu Ming 2. 2010. *Il sentiero degli dei*. Portogruaro: Ediciclo.

———. 2016. *Il sentiero luminoso*. Portogruaro: Ediciclo.

———. 2017. "Piccolo tour del disastro della Pianura Padana." Internazionale.it, May 15, 2017. https://www.internazionale.it/reportage/wu-ming-2/2017/05/15/tour-disastro-pianura-padana.

Yaeger, Patricia. 2011. "Editor's Column: Literature in the Ages of Wood, Tallow, Coal, Whale Oil, Gasoline, Atomic Power, and Other Energy Sources." *PMLA* 126 (2): 305–10.

Zanzotto, Andrea. 2011a. *Dietro il paesaggio*. In *Tutte le poesie*, 7–75. Milan: Mondadori.

———. 2011b. "Ormai." In *Tutte le poesie*, 12. Milan: Mondadori.

Zapf, Hubert. 2010a. "Cultural Ecology, Postmodernism, and Literary Knowledge." In *Redefining Modernism and Postmodernism*, edited by Hubert Zapf and Şebnem Toplu, 2–15. Newcastle upon Tyne: Cambridge Scholars.

———. 2010b. Preface to *Redefining Modernism and Postmodernism*, edited by Hubert Zapf and Şebnem Toplu, xi–xii. Newcastle upon Tyne: Cambridge Scholars.

———. 2016. *Literature as Cultural Ecology: Sustainable Texts*. London: Bloomsbury.

Zinato, Emanuele. 2012. "Alessandra Sarchi, *Violazione*." Allegoria. Accessed April 2020. https://www.allegoriaonline.it/index.php/i-numeri-precedenti/allegoria-n65-66/82-tremila-battute/659/524-alessandra-sarchi-qviolazioneq.

Index

adjacencies, 10
Adorno, Salvatore, "Petrochemical Modernity in Sicily," 56–57
aeropainting, 31, 207n19
"aeropoema del Golfo della Spezia, L'" (The aeropoem of the Gulf of La Spezia; Marinetti), 26–27, 207n19
aeropoetry, 31, 207n19
Aeschylus, 72
Africa, 30, 103, 110, 209n7
Agamben, Giorgio, 9
Agrigento, Italy, 20
Alaimo, Stacy, 6–7, 44, 101, 125, 128, 146, 148
    Bodily Natures, 133
All That Is Solid Melts into Air (Berman), 15–16
aluminum, 14
Aluminum Dreams (Sheller), 14
Alvaro, Corrado, *Itinerario italiano*, 195
Amberson, Deborah, 19–20, 67
Amianto: Una storia operaia (Asbestos: A blue-collar story; Prunetti), 11, 145–49, 150, 202–3, 219nn20–21
androcentrism, 28
animals, 103–4, 106–7, 110, 167–69, 184, 207n19
    death of, 181, 198
    devaluation of, 8
    pollution and, 81
    See also fish
Anthropocene epoch, 7, 59, 96, 158, 181, 202
anthropocentrism, 2, 8, 51, 157, 176
    questioning of, 7, 9, 33, 105, 201, 202
anthropomorphism, 2, 109, 127
anticittà, L' (Anticity; Boeri), 169–70
antifeminism, 54, 55
"Apua mater" (Ceccardi), 96–102
Apuan Alps, 11, 91, 95–96, 111, 116–17, 121, 123, 162, 202
Apuan Industrial Zone. *See* Zona Industriale Apuana
Apuan region, Italy, 92, 94, 120
architecture, 38–40, 61, 195, 208n25
Armiero, Marco, 5, 53, 125, 146, 147, 209n6
Arminio, Franco, 171

asbestos, 11, 124, 141, 142, 146–47, 152–53
    composition of, 218–19n17
    dangers of, 143–45, 149, 189, 202, 219n19
    uses of, 150–51
asphalt, 156, 162, 187, 189, 198, 199–200, 220–21n3
    environmental impact of, 193, 203
Atzeni, Daniele, *morti di Alos, I*, 211n37
"A un poeta nemico" (To a hostile poet; Quasimodo), 72–73
autofiction, 222n22
automobiles. *See* vehicles: cars
Avallone, Silvia, 218n7
    Swimming to Elba, 11, 126–27, 128–34, 135–41
avant-garde, 10, 25, 50, 205n10, 208n24

Baldanzi, Simona, 12
    Mugello sottosopra, 187, 188
Balla, Giacomo, 34
Baltimore, Maryland, 157
Banita, Georgiana, 58
Barad, Karen, 8, 88, 193, 213n3
Barca, Stefania, 94–95
Barron, Patrick, 172
Battaglia, Rosi, 150, 219–20n25
Bauman, Zygmunt, 192
Belloni, Eleonora, 210n16
Ben-Ghiat, Ruth, 215n19
Benjamin, Walter, 20, 21, 24
Bennett, Jane, 8, 35, 131, 139, 162, 201, 208n23, 218n14
    Vibrant Matter, 2, 34, 127–28
Berman, Marshall, *All That Is Solid Melts into Air*, 15–16
Bernazzoli, Dario, 51–52
Bertolucci, Attilio, 59
Bertolucci, Bernardo, *via del petrolio, La*, 71, 72
Bevilacqua, Mirko, 41
Bevilacqua, Piero, 6, 41
biofuels, 159
biopolitics, 9, 146, 147, 207n17
Blum, Cinzia Sartini, 30, 207n18
    Other Modernism, The, 27

Bodily Natures (Alaimo), 133
body, 35, 88, 105, 135, 174
   in futurism, 33–34, 45
   sexuality and, 133
body politics, 148
Boeri, Stefano, 155–56
   anticittà, L', 169–70
Bollate, Italy, 160
Bologna, Italy, 171, 172, 180
Bond, Emma, 19, 206n4
bone, 18, 22–23, 174
Bonfanti, Giovanni, *12 Dicembre*, 93
Bonifazio, Paola, 57
Bontempelli, Massimo, 210n20
   "522", 62–64
Braidotti, Rosi, 8, 134–35, 184
   Posthuman, The, 44
Brambilla, Michele, 149–50
   Morire d'amianto, 150
Brandelli d'Italia (Shreds of Italy; Cederna), 160
Breakthrough Institute, 157–58
Breu, Christopher, 224n2
brick, 75, 155, 162, 163, 179–80, 187
bronze, 91
Brown, Bill, 15
Buzzi, Paolo, 31

Calabria region, Italy, 12, 187, 191–93, 194, 195–96, 203
Calvino, Italo, 12, 204, 219n18
   "fabbrica nella montagna, La," 143
Campbell, Timothy, 207n17
Canavan, Gerry, 52
Cannamela, Danila, *Quiet Avant-Garde, The*, 35
capitalism, 148, 157
   global, 101, 140
   industrial, 23, 96, 146
Caproni, Giorgio, 126
carbon, 180–81
Carducci, Giosuè, 125
   "In riva al mare," 217n3
Carpathian Mountains, 174
cars. *See under* vehicles
Casale Monferrato, Italy, 11, 149, 150–51, 152–53, 203
Casid, Jill, 153
Cattabriga, Giovanni. *See* Wu Ming 2
Cattaneo, Giorgio, 111
Ceccagnoli, Patrizio, 208n23
Ceccardi, Ceccardo Roccatagliata, 213n6, 214n8
   "Apua mater," 96–102
   Dai paesi dell'anarchia, 98–102
Cederna, Antonio, 186
   Brandelli d'Italia, 160
Celati, Gianni, 12
cement, 160, 179, 190, 193, 199–200, 220–21n3, 223n35
Cemento S.p.A., 192
Cemento S.p.A. report, 190

"centauri, I" (The centaurs; Davì), 64–65
"522": *Racconto di una giornata* ("522": Tale of a day; Bontempelli), 62–64
climate change, 20, 205n4. *See also* global warming
Clockwork (von Bismarck and Charrière), 154–55, 162
coal, 6, 20, 136
coextensiveness, 50, 54, 91, 140
Cohen, Jeffrey, 19, 91
   Elemental Ecocriticism, 18–19, 203
Colombo, Luigi (pseud. Fillìa), 30, 53
colonizing, 118, 119, 123, 209n7
concrete, 154–56, 162, 165–66, 173, 179–80, 187–89, 190, 200, 203
Consolo, Vincenzo, *olivo e l'olivastro, L'*, 70–71
continuity, 33–34, 35
contro in testa: Gente di marmo e d' anarchia, Il (Resistance in the head: Marble people and anarchists; Rovelli), 93, 115–20, 123
Corneli, Franca Maria, "Creative Anxiety of the Hydroelectric Plant Nera Velino, The," 209n4
corporeality, 10, 35, 40, 45, 130–31
counterutopia, 162
"Creative Anxiety of the Hydroelectric Plant Nera Velino, The" (Corneli), 209n4
Cresswell, Tim, 192
   On the Move, 187
cultural ecosystem, 45, 49, 82
Curi, Fausto, 208n24
Curino, Laura, "Malapolvere," 151–52

Dai paesi dell'anarchia (From the villages of anarchy; Ceccardi), 98–102, 116
Danna, Daniela, 158
D'Annunzio, Gabriele, 125–26
Davì, Luigi, "centauri, I," 64–65
DDT (dichlorodiphenyltrichloroethane), 178
death, 135–36, 145–46, 150, 166–67, 188, 192
De Boever, Arne, 146
dehumanization, 213n48
DeLanda, Manuel, 8, 174, 175
Deleuze, Gilles, 10, 123
Delville, Michel, 208n24
De Martino, Ernesto, 185
De Michele, Girolamo, 146
demolition, 144, 145, 164
Depero, Fortunato, 34
Dessì, Giuseppe, "fidanzato, Il," 65–66
dialectical images, 20, 21
Dietro il paesaggio (Behind the landscape; Zanzotto), 171–72
dioxin, 141–42, 143, 189, 205n7, 218n16
disaster narratives, 24
disasters, 162, 183
   environmental, 6, 118–19, 120–21, 150, 205n7
   natural, 109, 110–11, 149–50
disruption, 6
12 Dicembre (12 December; Pasolini and Giovanni), 93

Donati, Riccardo, 222n28
donna e il futurismo, La (Woman and futurism; Goretti), 54
Donnarumma, Raffaele, 17
Duckert, Lowell, *Elemental Ecocriticism*, 18–19
dystopia, 11, 50, 155, 156, 161, 169

ecocrime, 150
ecocriticism, 10, 27, 203–4
  functions of, 12
  material, 7, 33–34, 49, 207–8n20
  postmodern, 32
Ecocriticism and Italy (Iovino), 57
ecofeminism, 28–29
Ecology of Modernism: American Environments and Avant-Garde Poetics, The (Schuster), 49
ecomaterialism, 5, 7, 9, 25, 201, 215n17, 223–24n39
ecomodernism, 158
econarratology, 7
ecosemiotics, 173, 176
Ecosickness in Contemporary U.S. Fiction (Houser), 78–79
Egypt, 103, 107, 110
EJAtlas. *See* Environmental Justice Atlas
Elba, Italy, 6, 141
electricity, 13, 51, 52–53, 65–66
Elemental Ecocriticism (ed. Cohen and Duckert), 18–19, 203
elements, 1, 18–19
Elias, Norbert, *Society of Individuals*, 223–24n39
embodied knowledge, 74
embodiment, 35, 208n23
Emilia-Romagna region, Italy, 172
"End of Sand, The" (Owen), 202
ENI. *See* Ente Nazionale Idrocarburi
entanglement, 9, 11, 24, 106, 131, 193, 201
  with landscapes, 17, 19
  narratives of, 3
Ente Nazionale Idrocarburi (ENI; Italian Hydrocarbon Corporation), 56, 57–59, 64, 73, 118
  films of, 71–72, 210–11n24
  publications of, 11, 59–60, 210n16
"Environmental Humanities and the Futures of the Human, The" (Heise), 157–58
environmentalism, 192
environmental justice, 80, 162, 189
Environmental Justice Atlas (EJAtlas), 162, 221n7
Environment and Behavior (von Uexküll), 179
E-PRTR. *See* European Pollutant Release and Transfer Register
Esposito, Roberto, 9
Esso Oil Company, 51–52
Eternit, 149, 150, 151
Ethiopia, 54, 209n7
European Pollutant Release and Transfer Register (E-PRTR), 78
Expo 2015, Milan, Italy, 11, 158–64, 169, 170, 199–200, 221n8

expressivity, 2, 5, 6
extinction, 199, 201, 204
extraction, 20, 23, 203
extractivism, 22, 206n11

fabbrica del panico, La (The panic factory; Valenti), 146
"fabbrica nella montagna, La" (The factory in the mountain; Calvino), 143
factories, 131, 133, 135–38, 147, 218n6, 218nn10–11
Falci, Eric, 16, 206n2
Falco, Giorgio, 11, 164–69, 221n10, 221n13
Farina, Almo, 182–83, 223n32
Farmoplant chemical plant, Massa Carrara, Italy, 6, 118–19, 120–21, 205n7
fascino dell'Egitto, Il (Egyptian fascination; Marinetti), 27, 35
fascism, 54, 55, 118
Fava, Giuseppe, 151, 220n26
feminism, 7, 148
feminization, of nature, 26, 28, 54
Ferrara, Giuseppe, 212n39
  Gela antica e nuova, 66, 72–77
Ferroni, Giulio, 222n22
Fiat Automobiles, 62
fiber cement, 149
"fidanzato, Il" (The fiancée; Dessì), 65–66
Fillìa. *See* Colombo, Luigi
fire, 20, 75, 76
fish, 81, 168–69, 198–99, 224n45
floods, 108–9, 110–11
"Fondazione e manifesto del futurismo" (Foundation and manifesto of futurism; Marinetti), 50
food, 30–31, 81, 158–59, 181
fossil fuels, 20–21, 47, 48, 49, 60. *See also* coal; oil; petroleum
Foucault, Michel, 207n17
fragmentation, 20, 108, 155–56, 221n9
Fuel (Pinkus), 76
"fumo, Il" (The smoke; Pirandello), 17, 20–24
futurism, 15, 17, 25–28, 50–51
  gender and, 54–55
  landscapes and, 30, 31–32
  language and, 208n24
  matter and, 32–35
  nature and, 40–41, 45, 53–54
Futurist Cookbook, The (Marinetti and Colombo), 30–31, 53

Gadda, Carlo Emilio
  meraviglie d'Italia, Le, 66
  "pozzo n. 14, Il," 66–69
Galanti, Carlo Mazza, 116
Garfagnana region, Italy, 111
gasoline, 50, 51, 52
Gatto Selvatico, Il (ENI), 11, 59–62, 64, 66

Gela antica e nuova a.k.a. *I miti e il petrolio* (Gela ancient and new a.k.a. The myths and petroleum; Sciascia and Ferrara), 66, 71, 72–77
"Gela: Realtà e condizione umana" (Gela: Reality and human condition; Sciascia), 66, 69–70
geocoreography, 91, 213n1
geocriticism, 7
Ghosh, Amitav, "Petrofiction," 46–47
Giannini, Sirio, 213n5
Gli iconemi: Storia e memoria del paesaggio (Iconemi: History and memory of landscape; Turri and Jodice), 156
Global South, 4, 5
global warming, 20, 22, 163
Glotfelty, Cheryll, 5
Goretti, Maria
  donna e il futurismo, La, 54
  "Song of Petroleum," 54–56
grande Milano tradizionale e futurista, La (The great traditional and futurist Milan; Marinetti), 29
Gregson, Nicky, 144
Guarnieri, Silvio, 106
Guattari, Félix, 10, 123
"guerra elettrica, La" (The electrical war; Marinetti), 50–51, 52

Hall, Marcus, 5
Haraway, Donna, 8
Härmänmaa, Marja, 26
Harvey, David, 160, 165
  Spaces of Hope, 156–57
Heise, Ursula, 102, 156
  "Environmental Humanities and the Futures of the Human, The," 157–58
  Sense of Place and Sense of Planet, 194
Hekman, Susan, 44
Houser, Heather, *Ecosickness in Contemporary U.S. Fiction*, 78–79
humanism, 9, 157
humanities, 3
  environmental, 7, 10, 12, 158
hybridity, 74, 106, 197

Iengo, Ilenia, 125, 146, 147
incorporation, 30–31
industrialization, 21, 23, 56, 69, 94–95, 117, 203, 213n48
Inhuman Nature (ed. Cohen), 91
"In riva al mare" (On the shore of the sea; Carducci), 217n3
interconnectedness, 4, 101
interdependence, 47, 49, 165–66, 201
interpenetration, 18, 100
Iovino, Serenella, 3, 4, 5, 7, 74, 76, 140, 205nn4–5, 218n13
  Ecocriticism and Italy, 57
iron, 6, 131, 132, 141, 188
ISLE, 187

isomorphism, 2
Italcementi, 161
Italian Hydrocarbon Corporation. *See* Ente Nazionale Idrocarburi
Itinerario italiano (Italian itinerary; Alvaro), 195

Jackson, B. John, 221n8
James, Erin, 4–5, 205n6
Jansen, Monica, 145–46, 151, 219n21
Jodice, Mimmo, *Gli iconemi*, 156

Kanafani, Ghassan, *Men in the Sun*, 22
Karst landscape, 18–20
Kerényi, Karol, 210n22
Kern, Robert, 12
Krumm, Ermanno, 2

landscapes, 19, 53, 94, 115, 154, 203
  corporeal, 189
  cultural, 83
  degradation of, 127, 188
  evolutionary, 140
  feminized, 54–55
  futurism and, 26, 29–30, 31–32
  industrial, 94–95, 99–100
  language of, 181
  official, 210n11
  pollution of, 87–88
  private, 182–83
  relations to, 5–6, 43, 76–77, 112, 171
  rural, 180
  as texts, 98
  as vengeful, 100–101
  violence toward, 121, 185
landslides, 183
language, 177, 181, 184, 208n24
  vernacular, 148–49, 212n41
Latour, Bruno, 8
lead, 6, 148
Lehtimäki, Markku, 212n34
LeMenager, Stephanie, 66
  Living Oil, 47
Leopardi, Giacomo, *Moral Essays, The*, 173–74
Leopold, Aldo, 19
Levi, Primo, 6, 205n2
  Other People's Trades, 12–13
  Periodic Table, The, 1–3, 142–43
Liguria region, Italy, 11
Living Oil: Petroleum Culture in the American Century (LeMenager), 47
Livorno, Italy, 126
"logica di Ahmed Bey, La" (The logic of Ahmed Bey; Marinetti), 28
Lombardy region, Italy, 156, 160–61, 164
Lopez, Barry, 123, 124, 213n4

Maccari, Mino, 59
Macdonald, Graeme, 47

machines, 31, 44, 50, 51, 155, 207n19
  nature and, 26, 29, 218n8
  See also vehicles
Mafarka the Futurist (Marinetti), 26, 32, 54–55
Maggiani, Maurizio, 96
  Meccanica celeste, 111–15
Magnaghi, Alberto, 159, 220n2
Malapolvere: Una città si ribella ai "signori" dell'amianto (Evil dust: A city rebels against the asbestos "masters"; Mossano), 150, 151
"Malapolvere: Veleni e antidoti per l'invisibile" (Evil dust: Poisons and antidotes to the invisible; Curino), 151–52
Malm, Andreas, 20, 22
Mangano, Daniel, 31, 207n19
Manifesto dell'architettura futurista (Manifesto of futurist architecture; Sant'Elia), 208n25
Mao, Douglas, *Solid Objects*, 16
Maran, Timo, 172, 173
marble, 11, 94–96, 100–101, 108–10, 111–12, 202
  dangers of, 93, 214n11
  physical properties, 92, 116
  uses of, 91, 114–15, 123, 161, 217n27
Marchesini, Roberto, 9, 106, 199
Marcuse, Herbert, 203
Marinetti, F. T., 10, 17, 24–45, 27–28, 33, 34, 45, 103, 207n15, 208n23
  "aeropoema del Golfo della Spezia, L," 26–27, 207n19
  cities and, 30, 209n9
  fascino dell'Egitto, Il, 27, 35
  "Fondazione e manifesto del futurismo," 50
  grande Milano tradizionale e futurista, La, 29
  "guerra elettrica, La," 50–51, 52
  landscapes and, 26, 29–30, 31
  "logica di Ahmed Bey, La," 28
  Mafarka the Futurist, 26, 32, 54–55
  "Meandri di un rio nella foresta brasiliana," 27
  monoplan du pape, Le, 41, 43
  Novelle colle labbra tinte, 27–28
  "nuova religione-morale della velocità, La," 51
  "poema non umano dei tecnicismi, Il," 27, 35, 37–38
  Ricostruire l'Italia con architettura futurista Sant'Elia, 38–40
  Roi Bombance, Le, 41
  Spagna veloce e toro futurista, 27
  teatrino dell'amore, Il, 36–37
  Una sensibilità italiana nata in Egitto, 29
  "11 baci a Rosa di Belgrado," 28
  Untameables, The, 27
  Venezianella e Studentaccio, 208n25
  Vengono, 36
  Vulcano, 40–45
Masala, Francesco, 212n41
  parroco di Arasolè, Il, 79, 82–89
Massa Carrara, Italy, 6, 94, 95–96

material   249
  agentic, 6, 93–94
  corporeality of, 45
  interaction with, 2, 5
materialism, 7
  vital, 34, 45
materiality, 68
  agentic, 124
  transformation of, 212n40
  vibrant, 8
  vital, 131, 139
material memoirs, 11, 146, 148
Mattei, Enrico, 58–60
matter
  agentic, 7, 82, 162
  devaluation of, 8
  futurism and, 32–35
  nature displaced by, 26
  primal, 107
  as text, 3, 207–8n20
  textual, 208n24
  vibrant, 133, 136, 162, 208n23
  vitality of, 37
Mazzotta, Massimiliano, *Oil*, 79–82, 88
McLeod, Neal, 85
"Meandri di un rio nella foresta brasiliana" (Meanders of a stream in the Brazilian forest; Marinetti), 27
Meccanica celeste (Heavenly mechanics; Maggiani), 111–15
memory
  collective, 151
  material, 125, 148, 149
  of place, 198, 199
Men in the Sun (Kanafani), 22
meraviglie d'Italia, Le (The wonders of Italy; Gadda), 66
Merleau-Ponty, Maurice, 42
metal, 128, 131–32, 143, 204. *See also* iron; steel
Mezzadra, Sandro, 22, 23
Mignolo, Walter, 159
Milan, Italy, 29, 160–61
Milani, Giulio, 96
  terra bianca, La, 93, 115–16, 120–23
mindscape, 11, 205n12
Minervino, Mauro F., *Statale 18*, 12, 187, 192–200
mines, 6, 20, 21–22, 142–43, 202, 206n9
mio Carso, Il (My Karst; Slataper), 10, 18–20
misogyny, 54
mobility, 170, 187, 191, 192, 196
modernism, 10, 27, 51, 205n10
  elemental, 16
  Italian, 14–17, 24, 102–3, 104
modernity, 5, 47, 101, 119
modernization, 15, 17, 26, 79
monoplan du pape, Le (The pope's monoplane; Marinetti), 41, 43
Montedison, 56, 57, 118
Moral Essays, The (Leopardi), 173–74

Morire d'amianto: Il caso Eternit; La fabbrica, le vittime, la giustizia (Dying by asbestos: The Eternit case; Plant, victims, justice; Mossano and Brambilla), 150
morti di Alos, I (The dead of Alos; Atzeni), 211n37
Morton, Timothy, 133, 202
Moscardino (Pea), 103
Mossano, Silvana, 149–50
  Malapolvere, 150, 151
  Morire d'amianto, 150
Mount Etna, 40, 41, 43
Mugello region, Italy, 187, 188–92, 203, 223n36
Mugello sottosopra: Tute arancioni nei cantieri delle grandi opere (Mugello upside down: Orange overalls in the building sites of the great works; Baldanzi), 187, 188
Munif, Abdelrahman, *Trench, The*, 47
Mussolini, Arnaldo, 14
Mussolini, Benito, 209n5

Naples, Italy, 6, 193
nature
  agency of, 7, 173
  conquest of, 26
  culture and, 2, 6
  as feminine, 26, 28
  futurism and, 40–41, 45, 53–54
  as hero, 27
  language of, 133
  as maternal, 26
  returns to, 16, 215n19
necroregions, 138–39, 140, 218n13
Negarestani, Reza, 211n34
Neilson, Brett, 22, 23
New Italian Epic (NIE), 112
New Italian Thought, 8–9
Newman, Richard, 116–17, 125
new materialism, 6, 7–8, 9
NIE. *See* New Italian Epic
Nixon, Rob, 102, 121, 143, 210n11
nomadic metaphysics, 199, 224n40
nonanthropocentrism, 9, 35, 169, 174
nonlinearity, 8
nonplaces, 68, 155, 180, 222–23n30
No TAV movement, 188
Novelle colle labbra tinte (Stories with painted lips; Marinetti), 27–28
Nozzoli, Anna, 55
"nuova religione-morale della velocità, La" (The new religion-morality of speed; Marinetti), 51

objectification, 2, 129–30, 203
oil, 20–21, 46, 49, 60, 74, 88–89, 209n7, 212n34
  futurism and, 49–50, 51–52
  industrialization and, 69
  Italian economy and, 47–48, 49, 51, 56–58, 161, 209n2

oil encounter, 46–47, 49
Oil: La forza devastante del petrolio; La dignità del popolo sardo (Oil: The devastating force of petroleum; The dignity of Sardinians; Mazzotta), 79–82, 88
Oliveira, Henrique, *Transcorredor*, 170–71
olivo e l'olivastro, L' (The olive tree and the oleaster; Consolo), 70–71
One, No One, and One Hundred Thousand (Pirandello), 108, 207n14
On the Move: Mobility in the Modern Western World (Cresswell), 187
Oppermann, Serpil, 3, 7, 205n4
"other," 109, 173
Other Modernism, The (Blum), 27
Other People's Trades (Levi), 12–13
overbuilding, 155, 187, 195
Owen, David, "End of Sand, The," 202

Paesaggio, costituzione, cemento (Landscape, constitution, cement; Settis), 185, 193
paesaggio come teatro, Il (Landscape as theater; Turri), 76
Paradise Built in Hell, A (Solnit), 162
Parma, Italy, 51
parroco di Arasolè, Il (The parson of Arasolè; Masala), 79, 82–89
Pasolini, Pier Paolo, 12, 82
  12 Dicembre, 93
  Petrolio, 59
Past, Elena, 5, 67, 74, 211n33
pastoral, 62, 94, 96, 210n19
Pattarozzi, Gaetano, 31
Pea, Enrico, 96, 102, 215nn14–16, 216n23
  Vita in Egitto, 103
  volto santo, Il, 104–11
"pedone al Motèl, Un" (A walker at the motel; Squarcia), 61–63
Penone, Giuseppe, 91–92
Periodic Table, The (Levi), 1–3, 142–43
Perniola, Mario, 215–16n21
Pero, Italy, 160
"Petrochemical Modernity in Sicily" (Adorno), 56–57
petrochemical plants, 56–57, 67, 72, 73, 78
petrocultures, 46, 49
petrofiction, 47, 59
"Petrofiction: The Oil Encounter and the Novel" (Ghosh), 46–47
petroleum, 10–11, 55, 58, 73–76, 202, 211n34
Petrolio (Petroleum; Pasolini), 59
petropastoral, 62, 64
petro-texts, 11, 49
Pickering, Andrew, 128
Piedmont region, Italy, 119, 149
Pinkus, Karen, 48, 59, 210n21
  Fuel, 76
Piombino, Italy, 6, 11, 126–27, 140–41, 218n6

Piovene, Guido, *Viaggio in Italia*, 95–96
Pirandello, Luigi, 10, 17–18, 202, 206–7n12, 215–16nn21–22
"fumo, Il," 20–24
One, No One, and One Hundred Thousand, 108, 207n14
plants, 8, 106–7, 112–13, 152–53, 173. See also trees
Plumwood, Val, 28
PMLA, 6
"poema non umano dei tecnicismi, Il" (The nonhuman poem of technicalities; Marinetti), 27, 35, 37–38
Poggi, Christine, 26, 32
Politische Geographie (Ratzel), 214n10
pollution, 84–86, 118–19, 165, 190, 217–18n5
 dangers of, 78, 80–81, 117, 126
Ponzanesi, Sandra, 205n6
postcolonialism, 4, 205n6
Posthuman, The (Braidotti), 44
posthumanism, 7, 8, 33, 135, 202
postmodernism, 14–15
Pound, Ezra, 16, 215n16
Po Valley, 12, 156, 164
"pozzo n. 14, Il" (The well n. 14; Gadda), 66–69
primordial liquid, 92, 138–39, 174
propaganda, 66, 72, 77
Provence region, France, 29
Prunetti, Alberto, 219n19
 Amianto, 11, 145–49, 150, 202–3, 219nn20–21
Purple Cloud, The (Shiel), 115

quarries, 91, 93, 98–101, 116, 117–18
 tourism and, 94–95
quartz, 180
Quasimodo, Salvatore, "A un poeta nemico," 72–73
Quiet Avant-Garde, The (Cannamela), 35

Raine, Anne, 27, 206n1
Ratzel, Friedrich, *Politische Geographie*, 214n10
Re, Lucia, 55, 103
Reactive Citizens (Cittadini Reattivi), 150, 219–20n25
Reggio Calabria, Italy, 193
reinhabitation, 140, 141, 153
Renaissance, 8
renewable energy, 160
Resilient Stories (Storie Resilienti), 150
Ricostruire l'Italia con architettura futurista Sant'Elia (Reconstructing Italy with Sant'Elia's futurist architecture; Marinetti), 38–40
Risi, Dino, 126
rivers, 27–28, 108–9, 152, 181, 209n4
Roi Bombance, Le (Marinetti), 41
Rome, Italy, 61–62
Rosa, Alberto Asor, 93
Rosignano Solvay, Italy, 126, 127, 149, 217–18n5

Rovelli, Marco, 92, 96
 contro in testa, Il, 93, 115–20
Rovina (Ruin; Vinci), 165
Ruhr Valley, Germany, 165
Rumianca poisons factory, 122

Saban, Analia, 92
sand, 202, 222–23n30
sandstone, 179–80
Sannazaro de Burgondi, Italy, 210–11n24
San Rocco, Italy. See Sarroch, Italy
Sant'Elia, Antonio, *Manifesto dell'architettura futurista*, 208n25
Sanzin, Bruno, 31
Saras (Società Anonima Raffinerie Sarde), 78, 79–82
Sarchi, Alessandra, *Violazione*, 12, 171, 172–85
Sardinia, 6, 77, 78–79, 84, 202
Sarroch (San Rocco), Italy, 78, 79–82, 213n48
Saviano, Roberto, 216n25
Schliephake, Christopher, 79
Schnapp, Jeffrey, 32
Schoonover, Karl, 212n40
Schuster, Joshua, *Ecology of Modernism, The*, 49
Sciascia, Leonardo
 "Gela," 66, 69–70
 Gela antica e nuova, 71, 72–77
 "Wine-Dark Sea, The," 77
Scott, James, 51
sedentarist metaphysics, 187, 192, 196, 199, 224n40
Seger, Monica, 5, 141, 218n16
selenite, 179–80
Semiologia del paesaggio italiano (Semiology of the Italian landscape; Turri), 156
Sense of Place and Sense of Planet (Heise), 194
sensibilità italiana nata in Egitto, Una (An Italian sensibility born in Egypt; Marinetti), 29
sentiero degli dei, Il (The gods' path; Wu Ming 2), 170, 188
sentiero luminoso, Il (The shining path; Wu Ming 2), 12, 162–64, 170
servitore del diavolo, Il (The devil's servant; Pea), 103
Settis, Salvatore, 160
 Paesaggio, costituzione, cemento, 185, 193
Seveso, Italy, 6, 165, 205n7
sexuality, 86–88, 133–34, 136, 208n24, 212n45, 222n26
Sheller, Mimi, *Aluminum Dreams*, 14
Shiel, M. P., *Purple Cloud, The*, 115
Sicily, 6, 17–18, 40, 57, 69–71, 202
Slataper, Scipio, 17, 206n4
 mio Carso, Il, 10, 18–20
Snyder, Gary, 150
social justice, 7, 101, 189
Society of Individuals (Elias), 223–24n39
socioenvironmental justice, 11

Solid Objects: Modernism and the Test of Production (Mao), 16
Solnit, Rebecca, *Paradise Built in Hell, A*, 162
Solvay, 126, 127, 149, 210n10, 217–18n5
Somigli, Luca, 108, 216n22
"Song of Petroleum" (Goretti), 54
Spaces of Hope (Harvey), 156–57
Spagna veloce e toro futurista (Speedy Spain and a futurist bull; Marinetti), 27
Squarcia, Francesco, "pedone al Motèl, Un," 61–63
Squier, Susan, 211n31
Statale 18 (State road 18; Minervino), 12, 187, 192–200
steel, 6, 124, 131–34, 137, 141, 161
steel mills, 127, 140, 218n6
stone, 91–92, 204
Stone: An Ecology of the Inhuman (Cohen), 91
"storied materiality," 144
"storied matter," 5, 94, 207–8n20
"Stories of Stone" (Cohen), 91
Strapaese movement, 104, 215n19
sulfur, 6, 20–25, 202, 204, 206nn8–10, 206–7n12
Sullivan, Heather, 49, 96, 210n19
sustainability, 158–59, 161, 187
sustainable economies, 159
Swimming to Elba (Avallone), 11, 126–27, 128–34, 135–41
Szeman, Imre, 47, 81

tar, 212n34
Taranto, Italy, 141
TAV. *See* Treno Alta Velocità
teatrino dell'amore, Il (The little theater of love; Marinetti), 36–37
Terni, Italy, 6
terra bianca: Marmo, chimica e altri disastri, La (The white land: Marble, chemistry and other disasters; Milani), 93, 115–16, 120–23
Terrosi, Roberto, 33
thanatopolitics, 146
thingness, 36, 128, 129
Thompson, Kara, 92, 188–89
Tomasi, Franco, 221n13
toxic autobiography, 116, 118, 120, 125
transcorporeality, 44, 49, 119, 125, 132, 152, 166, 167
Transcorredor (Oliveira), 170–71
transformation, 43, 73, 98, 115, 128, 144, 199, 212n40
trees, 53, 67, 207n14, 224n42
Trench, The (Munif), 47
Treno Alta Velocità (TAV), 12, 187, 188–89, 190–92, 221n5
Trevisan, Vitaliano, 155, 220n1, 221n9
Trieste, Italy, 18–19
Turri, Eugenio, 186, 187, 211n35
  Gli iconemi, 156
  paesaggio come teatro, Il, 76
  Semiologia del paesaggio italiano, 156

Tuscany region, Italy, 12, 98, 104
  cultural associations, 126
  industry in, 11, 122, 124, 190
  tourism and, 125
Tutter, Adele, 91
Tyrrhenian Sea, 125, 126, 193, 217n3, 224n44

ubicazione del bene, L' (The site of goodness; Falco), 164–69
"11 baci a Rosa di Belgrado" (11 kisses to Rose of Belgrade; Marinetti), 28
Ungaretti, Giuseppe, 102–3, 215n14
United States, 117, 218–19n17
Untameables, The (Marinetti), 27
Urban, Maria Bonaria, 151
urbanization, 26
utopia, 11, 22, 155, 156, 161, 162, 163–64, 169
  degenerate, 165
  futurist, 50, 52, 53
  negative, 188
utopianism, 156–58, 159

Val Bormida region, Italy, 119
Valdemarca, Gioia, 221n9
Valenti, Stefano, *fabbrica del panico, La*, 146
Valesio, Paolo, 33, 208n25
Variante Autostradale di Valico (VAV; A1 Highway Variant), 187, 189, 191
vehicles, 50, 64–65, 190
  cars, 29, 58, 61, 192, 195–97, 210n21, 222n21
  entrapment in, 196, 197
  personified, 62–64
Veneto region, Italy, 164
Venezianella e Studentaccio (Marinetti), 208n25
Vengono (They are coming; Marinetti), 36
Venice, Italy, 38, 40, 57, 209n9
Verdicchio, Pasquale, 5
Versilia region, Italy, 103, 104
via del petrolio, La (The way of petroleum; Bertolucci), 71, 72
Viaggio in Italia (Voyage in Italy; Piovene), 95–96
vibrancy, 11, 40, 133, 136
Vibrant Matter: A Political Ecology of Things (Bennett), 2, 34, 127–28
Vinci, Simona, *Rovina*, 165
Violazione (Violation; Sarchi), 12, 171, 172–85, 200
Vita in Egitto (Life in Egypt; Pea), 103–4
vitality, 37, 51, 107, 127
  of buildings, 40
  of landscapes, 20, 41
  materiality and, 35, 131
  performative, 144
volcanoes, 41, 42–44
volto santo, Il, 104–11
volto santo, Il (The holy face; Pea), 103
von Uexküll, Jakob, 178
  Environment and Behavior, 179

Vulcano: 8 sintesi incatenate (Volcano: 8 chained syntheses; Marinetti), 40–45

war, 50–51, 67, 121
   glorification of, 24, 26
   See also World War I; World War II
water, 27–28, 108–9, 162, 217n3
   hydropower and, 53
   pollution of, 81, 152
   See also floods; rivers
Wenzel, Jennifer, 47
West, Rebecca, 59
"Wine-Dark Sea, The" (Sciascia), 77
Withers, Jeremy, 192
wood, 161, 167
workers, 20, 21–22, 75, 98–99, 111–12, 206n9
   dangers to, 80, 93, 95–96, 100, 121, 145–46, 188, 189, 214n11
   exploitation of, 147

marginalized, 219n18
migrant, 191, 192, 223n38
World War I, 24, 48
World War II, 2, 115, 117, 118
   economy after, 6, 58, 171, 186, 211n28
Wu Ming collective, 112
Wu Ming 1, 112, 146, 219n20
Wu Ming 2, 11–12, 170–71, 221nn5–6
   sentiero degli dei, Il, 188
   sentiero luminoso, Il, 12, 162–64

Yaeger, Patricia, 6

Zanzotto, Andrea, *Dietro il paesaggio*, 171–72
Zapf, Hubert, 10, 15, 82, 115, 124
ZIA. *See* Zona Industriale Apuana
zinc, 6, 148
Zona Industriale Apuana (ZIA; Apuan Industrial Zone), 118, 121

www.ingramcontent.com/pod-product-compliance
Lightning Source LLC
Chambersburg PA
CBHW022046290426
44109CB00014B/1006